For Robert
from
Rumana
with love

York - September '06

Zohra

To Zeenuth Futehally's grandchildren,
the last generation to have known the author—
Nilam, Hashim, Ahmad, Naella, and Nikhat

Zohra

by
Zeenuth Futehally

Edited by
Rummana Futehally Denby

OXFORD
UNIVERSITY PRESS

OXFORD
UNIVERSITY PRESS

YMCA Library Building, Jai Singh Road, New Delhi 110001

Oxford University Press is a department of the University of Oxford. It furthers the
University's objective of excellence in research, scholarship, and education by
publishing worldwide in

Oxford New York
Auckland Bangkok Buenos Aires Cape Town Chennai
Dar es Salaam Delhi Hong Kong Istanbul Karachi Kolkata
Kuala Lumpur Madrid Melbourne Mexico City Mumbai Nairobi
São Paulo Shanghai Taipei Tokyo Toronto

Oxford is a registered trademark of Oxford University Press
in the UK and in certain other countries

Published in India
by Oxford University Press, New Delhi

© Oxford University Press 2004

The moral rights of the authors have been asserted
Database right Oxford University Press (maker)

First published in India 2004

All rights reserved. No part of this publication may be reproduced,
or transmitted, in any form or by any means, without the prior permission
in writing of Oxford University Press, or as expressly permitted by Law,
or under terms agreed with the appropriate reprographics rights organization.
Enquiries concerning reproduction outside the scope of the above should be
sent to the Rights Department, Oxford University Press, at the address above

You must not circulate this book in any other binding or cover and you must
impose this same condition on any acquirer

ISBN 0 19 566716 6

Typeset in Bembo
by Eleven Arts, Keshav Puram, Delhi 110 035
Printed by Pauls Press, Delhi 110 020
Published by Manzar Khan, Oxford University Press
YMCA Library Building, Jai Singh Road, New Delhi 110 001

Contents

Editor's Introduction vii

Note on the Text ix

Zohra 1

Glossary 256

Appendix 259

Editor's Introduction

Zohra is not merely a novel—it is also an intensely nostalgic historical record of a city and a social commentary on a unique way of life which Zeenuth Futehally 'felt an urgency to record, for owing to the passage of time it was fast disappearing'. It chronicles the life of a girl born in the early twentieth century into a wealthy and privileged Muslim family in Hyderabad, a city described as a 'replica of paradise itself'.

The founder of Hyderabad, Mohammed Quli Qutb Shah was the ruler of Golconda in the Deccan, one of the numerous princely kingdoms into which present-day India was formerly divided. Born in 1565, he was the son of a Hindu princess and a Muslim sultan and went on to become one of the finest poets of his age, writing not only in Persian and Urdu which were the court languages, but also in Dakhni, the language spoken by the common people. Thus, the fusion of Hindu and Muslim customs in Hyderabad produced a culture unlike any other in India and the author's deeply held social and political beliefs were founded on the religious tolerance and harmony which prevailed as a result.

The city's history dates back to the end of the sixteenth century when Golconda was becoming overcrowded and unhealthy. And so, according to the renowned poet Firishta, 'on an auspicious day when the Moon was in the constellation of Leo and Jupiter was in its own mansion', Quli Qutb Shah ordered the preparation of plans for a new city—Hyderabad—to be built across the Musi river, a city which would be unrivalled in the world.

The society in which Zohra was growing up in Hyderabad knew no flexibility. A girl of her background was expected to comply unreservedly with the wishes of her parents at the expense of her own dreams and desires.

The stifling mores of the day demanded that girls confine themselves to a secluded life in preparation for marriage. In such a world Zohra's talents for poetry and painting are suppressed in favour of the inevitable arranged marriage.

The bridegroom chosen for her is Bashir, a man her equal in social standing, brilliant but arrogant, who loves his wife with complete devotion but who shares none of her creativity and whose temperament prevents him from any demonstration of the tenderness he actually feels for her. Zohra is thus condemned to a solitude in which her life is circumscribed by domestic obligations.

Into this cultural desert enters her brother-in-law Hamid, very much the face of modern India, committed to justice and social equality and dedicated to the teachings of Mahatma Gandhi, living life on his own terms and in constant discord with the expectations of his family. Zohra who is gifted, sensitive, and passionate but has an inviolable sense of propriety and duty, is caught between these two men. Greatly frustrated and intensely vulnerable, she strives to subjugate her own needs in order to comply with the requirements of her husband and his family. The sacrifice she makes in order to stabilize her sister-in-law Safia's dysfunctional life is heartbreaking; her reward, suspicion and betrayal.

Zohra's personal struggle for self-expression is paralleled by India's growing desire for independence, which was spreading throughout the country. Mahatma Gandhi's new weapon of non-violence shattered this quietly cultured but stagnant Hyderabadi society, igniting a patriotic consciousness in a whole generation of young Muslim men and women and bringing them into conflict with their elders, a confrontation that would previously have been inconceivable.

Against this social and political background, Zohra is propelled into a world of turmoil and rapid social change. Her political awakening and emotional flowering become inextricably linked. But Zohra is trapped in a cage of social conventions which offers no release. The escape when it comes is unbearably painful—the tragedy overwhelming.

Note on the Text

I remember my mother's reaction to the publication of her novel *Zohra* (1951) as a mixture of exhilaration and regret: her inevitable elation was tempered by a profound disappointment at the lack of editing on the part of the publisher. Writing in a foreign language, and one which the author felt she 'could not handle with sufficient confidence' made, in her view, the task of sensitive editing all the more crucial; and she felt strongly that the novel had been compromised by its absence.

When I had occasion, therefore, to spend extended periods in India in 1986 and 1987, she and I had comprehensive discussions on what form the editing should have taken. Her thoughts were clear and her directions unambiguous. My detailed notes of those conversations have formed the basis for my revisions to this edition, which largely encompass the following:

One of the most significant blunders in the initial publication was that Chapters 4 and 5 had been switched around, and a few lines of text omitted at the beginning of each chapter. I have restored the original sequence and the missing text.

Another of Zeenuth Futehally's concerns was to minimize the distraction posed to the reader by the need to constantly refer to a glossary. As some of the words in the original glossary are either now in current English usage, or appear in the dictionary, they no longer need clarification and so have been dropped. In other cases, I have added a strategic word here or a sentence there, and, on one or two occasions, a short paragraph to incorporate into the text the meaning of various rituals and customs which otherwise would have made lengthy reading in the glossary. Where a political event could be elucidated by the insertion of a date, I have done so. Individually appropriate names have also been given to a host of minor characters, a bewildering number of who were referred to as either 'the Begum' or 'the Nawab', causing confusion.

And finally, I have removed or replaced repetitions where possible, and corrected obvious grammatical and spelling mistakes which should never have escaped even the most cursory editorial process. However, I have been acutely aware of the rhythm and idiom of local phraseology which gives the novel its integrity and authentic charm, and have made no attempt to compromise this by sterilizing the language in any way.

1

'*Ai-hai!*' exclaimed the *unnie*. 'What can we do with these girls of today?' The old nurse had a habit of using this recurring phrase with ever-fresh wonderment.

If the *unnie* had a name, it had long since been forgotten. She had been Zubaida Begum's nurse, and as such, held a very special position of trust and privilege in the household and was affectionately known by the entire family simply as Unnie.

Unnie now strolled into the high-walled garden and stopped to watch Zohra swinging heedlessly, eyes aglow and hair dishevelled, her lithe body moving rhythmically with the swing. Gulab, her maid, was standing behind it, giving her a push now and again.

Seeing her, Unnie placed her hands on her hips and cried: 'Allah, *masha-Allah*! Gulab, you aunt of Satan, what are you doing? God forbid—and may such a disaster never overtake even Chhoti Bibi's enemies—but do you want her to fall flat on her face?'

Zohra gradually slowed down.

'Oh Unnie, when will you stop scolding others for my misdeeds?'

'*Ai-hai,* Chhoti Bibi, may Allah preserve you for defending the poor, but this ....'

'I asked her to push the swing,' interrupted Zohra impatiently. Turning to Gulab, she gave her a conspiratorial smile; Gulab smirked.

Zohra continued: 'I am not a china doll which will fall and break in an instant. It is not so easy for a human being to die.' She laughed light-heartedly.

'*Towba,* Chhoti Bibi. What evil words to escape from a young girl. They should not even form themselves on your innocent lips.'

'Why Unnie, is it so awful to die?' asked Zohra teasingly. She came and put her arm around her.

At this, Unnie gave an exultant grunt, trying to make it sound reproving, and exclaimed: '*Owi*, everything is fun to the young people of today: Allah willing, you will live for hundreds and hundreds of years; may the bridegroom of your auspicious kismet arrive soon; and may the flowers of your bridal garland blossom swiftly and may ....'

'Is this a blessing?' cut short Zohra, embarrassed by the usual turn of Unnie's benedictions.

Unnie made no direct reply, but exclaimed excitedly: '*Owi*, what has happened to you all? In our days, girls behaved like girls.'

'Yes,' said Zohra with a toss of her head, her eyes bright with amusement, 'those good old days!' Then she musingly added, 'and what an ageless tune. It is a classic that will never grow stale. For, ever since the fall of Baba Adam and Bibi Havva—Adam and Eve—with each successive generation, we have been descending lower and lower into a decline. Is that not so?' She turned to Unnie and gave her a hug. Then, more seriously, 'I wonder what people would do if the good old days really came back.'

'*Owi*, Chhoti Bibi,' said Unnie chuckling, 'like departing friends, the old days only ever show their backs to us. May Allah preserve you for ever and ever! But in our days, girls never entered the outer garden, not even when it was high-walled. What is the zenana courtyard for? Your mother—may Allah bless her—never dreamed of doing such things when she was your age.'

'But Unnie, how fearfully dull life would be if daughters did exactly what their mothers had done before them, just like the recurring decimal forever and ever without any change.'

'Allah! Chhoti Bibi, you talk so glibly. What is all this studying for?' She sounded genuinely perplexed. 'What do you want to do with foreign books? Learning is so hard—like trying to chew steel nuggets. Were your grandmother alive—Allah bless her memory—she would never have allowed you to delve into such books. Is not our own knowledge enough for you? And as for your high spirits, she would have made you sit down with a tray of mixed rice and lentils and separate the grains. That was how girls were disciplined in our days. Your father the Nawab Sahib—may he live long—has spoilt you as if you were a son.' Unnie's laughter was full of fond disapproval.

With the satisfaction of having done her duty, she now took a stroll around the garden, plucking jasmine buds. Zohra and Gulab looked on with amusement, for they knew the old woman's weakness for adorning herself with flowers. Suddenly, she started screaming wildly.

'*Ai-hai*, you son of Satan. Get out at once, or else ....' Her curses were left half-uttered, for no sooner had she opened her mouth than the figure on the tree-top disappeared.

A voice from behind the wall interrupted loudly, 'I was only plucking tamarind.'

'Plucking tamarind!' Unnie shouted back at the top of her voice, then under her breath she murmured contemptuously. 'That may be so, and Allah forbid, at the same time also trying to pluck fruit with your glances from the neighbour's garden. For that, may dust be in your eyes blinding them.' Aloud, she cried, 'Your mouth will be filled with mire for telling such lies. If you wanted to pluck tamarind, could you not have given a warning first, as all decent people do?'

Zohra intervened quietly but firmly: 'Now Unnie, what is the sense in all this abuse? It was an oversight probably. How could he know that we ....'—she included Unnie and Gulab—'would be here now?'

'*Ai-hai*, Chhoti Bibi, you are unaware of their wily ways; of course it was with the intention of looking at you—may no evil eyes ever fall on you and yours—that he climbed that spreading tree. Otherwise, in what foreign land has he lived that he does not know a man must give warning before climbing up a tree or terrace overlooking another's garden? Come, Chhoti Bibi, come in; else, that man-child, who should be dust, may climb again.' She kept muttering to herself, only stray words of which were audible to the girls.

Annoyed, and not wishing Unnie to make further scenes, Zohra turned to Gulab and said: 'Let's go in.' Then significantly, 'It's getting too hot here.'

'I shall report to Begum Sahiba,' said Unnie pompously, as if about to perform an important task. Zohra wondered what fantastic tales Unnie would weave out of this simple episode. Passing through the corridor, they entered the zenana courtyard.

Zubaida Begum was seated on the divan stitching gold braid on to a pink net blouse that she was making for the trousseau of one of her maids. Unnie went straight to her.

Mehrunnissa, the Begum's elder daughter, stood near the fountain in the centre, scattering grain for the snow-white pigeons, which were hopping around and flying to and from their dovecotes carrying the grain to feed their young. Zohra, coming softly from behind in her slippers, called, 'Apa Jan!'

Mehrunnissa turned round with a start.

'*Towba*, Zohra, how you startled me; you came as lightly as dewdrops on green grass!' Pressing to her heart her tapering hands with glass bangles gleaming on her rounded wrists, she said affectionately, 'Allah, how it has started thumping.'

'Apa Jan, it's high time you learnt to control your heart better,' exclaimed Zohra, 'otherwise it may land you into trouble. Surely you didn't think that anyone who could really cause a flutter had entered the zenana?'

'*Ai-hai*, I don't know what you mean,' Mehrunnissa pursed her lips, conscious of her beauty that caused such a disturbance amongst the few boy-cousins privileged to see her.

Zohra did not reply. She knew nothing could detract from her sister's vanity. Slightly exasperated, she stood watching the pigeons.

Unnie's agitated voice talking to their mother drifted towards them. 'Let's go to Ammi Jan; I'm afraid of Unnie's exaggerations,' she said after a few moments, and before Mehrunnissa could ask any questions Zohra was hurrying away towards her mother. Mehrunnissa followed immediately.

As they approached, both Zubaida Begum's and Unnie's eyes turned to them. Zohra, slender and tall, walked swiftly but gracefully, with the eagerness of youth; her head, with wavy black hair woven into a plait, was set above a long shapely neck, which gave her a distinguished air. Mehrunnissa followed at a more leisurely pace, with a rhythmic swinging of her hips, adjusting her dupatta, smoothing her glossy black hair with both her hands, and looking voluptuously lovely.

They could hear Unnie excitedly relating the tale: '*Owi*, Begum, you must send word to the neighbouring house. I shall myself take the message to the Begum Sahiba there. May Allah preserve her, she is such a fine lady!' Unnie already relished the opportunity of carrying the exciting tale spiced with her own fantasy. 'But her son! *Owi*, does his behaviour become one of such a high and noble family?'

'How can you be sure that the boy was not really plucking tamarind?'

asked Zubaida Begum quietly. 'Anyhow, we should not draw unnecessary attention to the incident. A house full of young girls is always in a vulnerable position. If it ever happens again ....'

'*Owi*, Begum Sahiba, you are an angel; you can think evil of no one, but the way he was ogling ....'

'Unnie, Unnie,' broke in Zohra, 'you could not see that with your own eyes, but only with the help of your fertile imagination.'

Zohra had been forced impulsively to speak on a subject she would never ordinarily have referred to. She looked ill at ease.

Unnie knew it was no use trying to recount such incidents in front of Zohra, for they would inevitably lose colour. If only it were Mehrunnissa instead, what a rich yarn could have been woven around the incident. Only Unnie, like all those given to exaggeration, deceived herself into believing that she would then have spoken the truth. She now turned fondly to Mehrunnissa, chanting the stereotyped formula that she used when working up to a pitch of maternal benediction: 'May you live for ever and ever, and may your kismet be auspicious, and may I be fated to live to see that happy day when your sun-like bridegroom will come riding to fetch his moon-like bride!'

Mehrunnissa loved to hear this; the idea of marriage sent excited thrills through her, but she protested in feigned distaste: 'Oh, no! Unnie, don't say such things!'

With resentment against Zohra for having forced her to cut short her exciting story, Unnie complained: '*Owi*, Begum Sahiba, how can you allow Chhoti Bibi to go on with her studies? What is she going to do with all this? Look how thin she has become! She is as slender as a twig. Girls at her age should be plump and fresh.'

Zubaida Begum smiled a sad, benign smile.

Unnie, in her element, went on: 'Besides, I tell you this school is no place for our girls to go to. What can you expect when it is run by *mems*? As you know, ladies often come on the pretext of seeing the school; but *owi*! What have they to do with learning? They simply come to cast an eye over the girls and select brides for their sons. Only the other day there was a lady who visited the school, saying she wanted her daughter enrolled. Now, did I tell you about her? My head has turned so musty.' She scratched it, through her grizzled hair.

'I am sure you did, you bring such accounts often enough. You can never contain yourself even for an hour!' Zubaida Begum smiled indulgently.

Unnie, unmindful of it, continued as if the story were new: 'Now, this Begum went around the classrooms and lingered on to watch the girls when they were out in the playground. I, too, stood at a distance watching them, for I am always suspicious of these Begums, and *owi*! her eyes were constantly turning to our Chhoti Bibi—may Allah preserve her from evil eyes. I myself heard her enquiring of her niece as to who this girl was.'

'Unnie, Unnie!'

Zohra, who had been silently listening to Unnie's recital, now intervened in a tone which conveyed that she wanted this conversation to end. She could say nothing more.

'*Owi,* Chhoti Bibi, your mother must know all that is happening at school, else what am I here for? *Towba*! Is this the way to select brides for one's sons! Are they selecting—Allah forbid—vegetables, that they have to appraise them in this manner?' exclaimed Unnie, indignantly rolling her eyes.

'Oh, Unnie, stop it, please.'

Zohra found this gossip irksome. Mehrunnissa, however, was keenly interested in everything to do with the topic of marriage, and since this concerned Zohra and not herself, she could openly display her interest in the subject. Eagerly she asked: 'But Unnie, what sort of woman was she?'

Mehrunnissa, who had had her full share of the limelight when she had attended school, missed school only for occurrences such as these.

'*Ai-hai,* Mehrunnissa daughter!' her mother's tone was mildly reprimanding, 'these are not topics for the ears of young girls, let alone their tongues.'

Mehrunnissa was silenced, but Unnie continued: '*Owi,* Begum Sahiba, but what can they do when you send them to such a place? I hear girls talk and even amuse themselves with such prattle. *Ai-hai*! If you could know all the things that are happening. Why do they have a boys' school almost opposite? Some boys follow the girls' carriages on their bicycles. They say some shameless girls even encourage them by gazing out of the windows. Some say they even pass on notes to each other. What are girls coming to? It is for such fears that we poor folk discourage our girls from learning to read and write.'

'Now, enough, Unnie!' Zohra wanted to stop her, but Unnie continued unabashed.

'*Owi!* Chhoti Bibi, are there not boys often following our tonga also? It is only that I have a vigilant eye and keep them at their distance. *Ai-hai*, Begum Sahiba, I have to shout at them that, unless they go away, I shall have them taken to task and get them whipped by the coachman.' Unnie was full of her own zeal in warding off young men even from a closed horse-drawn carriage.

Zohra knew that such incidents sometimes did happen, but mostly they were blind dalliances, as the boys could never see her, and even she could only get a passing glimpse of them through the chinks of the curtains that veiled the windows of the tonga. She knew Unnie was exaggerating and making the unusual sound usual. Besides, how could she ever be certain that it was not mere coincidence. Maybe the boys were not following the tonga after all, but just returning home from school by the normal route. But Zohra could not discuss these matters before her elders, and it was only because of Unnie's complaints to their mother that the girls were able to listen in at all.

Mehrunnissa heard all this, and recalled her own school days, feeling a little wistful that she no longer attended school, and therefore was unable to play the heroine in such incidents. Her vanity had never left her in doubt that it was she who was being followed, even though she was not visible, for she was certain that fame of her beauty had reached everyone and that her carriage was deliberately singled out by the boys.

Zubaida Begum was naturally interested in Unnie's recital, but she refrained from asking any questions in the presence of her daughters.

2

The *musnud* had been spread in the centre of the zenana reception room. On this carpet, covered with a crisp white sheet and scattered brocade cushions, sat Nawab Safdar Yar Jung, smoking a handsome hookah of gunmetal intricately inlaid with silver. As he inhaled the perfumed smoke to the gentle gurgling of the water inside the bowl, his eyes quizzically rested on his wife seated a little apart. She looked a picture of beauty and repose. Zubaida Begum traced her ancestry back to Akbar and Babar, the Mughal emperors, her demeanour proclaiming her a scion of that proud dynasty. Looking at her, one had the impression of gazing at some old miniature painting of a bygone queen. And who can say what might have been, had the fortunes of that royal line not undergone such irrevocable changes? In the wake of the Mutiny of 1857, her grandfather had fled from Delhi, the capital of the Mughals. How well the Begum remembered her grandfather recounting to her the tales of those harrowing days: 'I was prepared to die serving my emperor, but his humane soul could not bear to see any more of this futile bloodshed. For, although not endowed with the fighting spirit of his great ancestors, he was an idealist, a poet. How well I remember Bahadur Shah, standing in the Red Fort and bidding farewell to his army: "The sword of Hind must now be sheathed," said he. It was then that my parents decided to flee from this beautiful city.'

With flaming eyes her grandfather would then demand of them: 'Could we have watched, with any equanimity, the English *firangis* walking through our marble palaces where flowed the *Nahr-i-Bihisht*, the Stream of Paradise, keeping the halls cool and fresh? Could we have tolerated a stranger delivering judgement under our scales of justice etched so beautifully on the arch of

Diwan-e-Aam, where the Emperor Shah Jehan first held public audience? And what about Amir Khusraw's immortal proclamation engraved on the archway of the Hall of Private Audience—*If there be a Paradise on Earth, it is this, it is this, it is this!* Could we have stood aside and watched Paradise turn to infernal hell? No, no! Our eyes had to turn away from these scenes of humiliation.' He would then recount how after long years of dangerous wandering they had ultimately come to Hyderabad and found shelter there. Even though the Nizam had declared himself an ally of the British, at least here, there was still some vestige of Muslim culture left.

Now Zubaida Begum remembered all this as she sat on the *musnud* at a respectful distance from her husband, skilfully embroidering a child's garment.

Although Zubaida Begum's once oval face was now a little too plump, it still retained its delicacy of features. The chin, though rounded, was firm; the nose was finely chiselled with a small bright diamond gleaming on the curve of the left nostril. The even teeth were stained with the juice of the betel nut, which she constantly made into a paan to chew. Above these features lay the shapely, but not too high, forehead, crowned with well-oiled and smoothly combed black hair, now sparsely streaked with silver. The large dark eyes, however, had a haunted look of suffering. Zubaida Begum had borne many children, but none except her two daughters had survived. Her fine lips, that once curved upwards as if in a perpetual smile, were now set in despondent sadness. Her body had lost the lissom elegance of her youth and it now spread amply on the cushions as she chided her husband.

'You must take heed, Nawab Sahib, of what I have said.' In spite of almost thirty years of marriage, husband and wife never addressed each other by their names. 'You must compel the manager to give you proper accounts. *Ai-hai*! he is ruining the estates. He is swindling us—sweetening his own mouth from every direction,' said the Begum, laying her sewing aside while she spoke.

'Begum, you will not let me trust anyone.'

'Nawab Sahib, it is with the help of people like you that such *badmashes* flourish. You must bring the accounts to me,' said the Begum with a firm gesture of her right hand. Zubaida Begum had strong opinions on how her family, her household, and the estate should be run, but little faith in her husband's judgement on any of these matters.

'Your Highness's command be upon my head and eyes,' said the Nawab Sahib, touching his head and eyes with mock gravity. He did not in the least mind his inability to balance his finances and willingly relied on his Begum to do so, accepting it all in good humour. He drew at his hookah again.

Zubaida Begum, used to his ways, resumed her sewing.

'Begum, for which fortunate infant are you making this?' asked Nawab Safdar Yar Jung by way of a pleasantry rather than a query.

'Tomorrow is the naming ceremony of Unnie's granddaughter, and I must finish this. It seems only the other day that I was sewing her daughter's trousseau, and in the twinkling of an eye here is the child.'

'Well, children will come and the world will go on, as sorrily as ever, with my Begum to brighten it by making little colourful garments!' exclaimed the Nawab Sahib, his manner lightly playful. But the Begum became instantly tense, and resting her sewing upon her lap, said: 'Is it not strange that all the women should think my hands auspicious? Because I have money they think everything I touch will turn to gold, and gold to them means happiness.' She gave a wistful laugh. 'In reality, I am the unfortunate one. It pleased Allah to bless me with children profusely, but it was writ in my kismet that so many should be recalled in their infancy.' Her voice was full of anguish. 'I, who have wept tears of blood! Allah knows how my arms have sought and my heart ached for each departed child!' Her eyes became moist and the Nawab Sahib was sorry for having unwittingly led her to this subject. The loss of their children pained him also, but there were other things to occupy a man.

'Begum, that was the will of Allah, or our kismet, call it what you will. Only, it was beyond our power. Let us now think of our daughters, who by His benign mercy, adorn our home.'

'Yes, and when I do think of them, you say I spoil them,' flashed back Zubaida Begum passionately. 'Allah knows, Mehrunnissa was a weak child; I have watched over her day and night. No change in her face, no turn of her body, went unnoticed. She is delicate like a rose petal. Only yesterday, she was complaining of palpitations. Allah be praised a thousand times that Zohra, at least, is stronger, although she is as thin as a bamboo twig. But she is overworked. When will you put an end to her studies?' Her eyes upbraided him as her voice sounded worried.

'Begum, her heart is set on books. You cannot snatch them away from her. Why mar her happiness? Adult lives are perforce dedicated to worry, whilst the joys of childhood and youth soon become a dream.'

'But some day she will have to marry and why sow discontent in her heart?' said Zubaida Begum who could not understand her husband's preoccupation with educating his daughters. 'After all this, she may never willingly settle down to domesticity. Allah forbid, but she has not to pass any doctorate, has she?'

'And why not, if that be her wish?' asked her husband. 'She is my son if Mehrunnissa is your daughter.' He gave a disappointed laugh signifying that the possibility of a doctorate was as far removed from his mind as it was from the Begum's. In Hyderabad, girls of the aristocracy had never followed professional careers. 'But anyhow,' he continued, 'Zohra is only sixteen. She has just finished her exam, and is now keen on literature and painting; surely there is no harm in that.'

'How often must I repeat that learned girls never settle down happily to domestic life,' rejoined Zubaida Begum irritably. 'They pick up ideas from reading unsuitable novels and always think marriage and children can be delayed.'

'You will see how well she settles down when we find the right bridegroom for her. By Allah, Begum, leave Zohra to me. It is sheer delight to read Persian with her. She might become a great scholar, even a poet. She is sensitive and has a delicate imagination. Yesterday, she composed some verses and they were supremely beautiful. She has a special gift for languages, she ....'

'But,' intervened the Begum, 'what is she going to do with all this? Surely, Nawab Sahib, she is not going to manage a home on such learning? If you like, let her continue with Urdu and Arabic. Of the rest, she has studied quite enough.' There was a finality in her tone.

'Begum, English is essential these days. Though I am no great English scholar, I know that it is the gateway to modern thought. All educated young men want their wives to speak it like the memsahibs do.' He made a humorous, disapproving face, for he was not a lover of English, the language of the new masters. 'As for Persian,' and at once his face lit up as if speaking of an old and constant beloved, 'it is the sweetest language in the world. There could be no poetry, no romance, no imagery, no delicacy of thought, without Persian. Persian gives out sweetness even as the rose does fragrance. Poetry and Persian are synonymous.' His eyes sparkled with fervour.

Softening, Zubaida Begum exclaimed with an indulgent smile: 'Allah save Zohra! You wish her to become a poet like her father!' A faint blush rose to her face as she recollected the romantic and sentimental poetry that her husband had written for her in the early days of their marriage.

'But why blame Zohra's poor father for all her shortcomings? What about your great grandmother? The Mughal princesses could express themselves in their own sweet language, but Zohra must now find a foreign tongue.'

Unwilling to agree with the Nawab Sahib that foreign learning was essential for Zohra, and saddened at the thought that she had not given her husband a living son, she was filled with compassion for him and, in a voice charged with suppressed emotion and urgency, she said: 'You must contract a second *nikah*, Nawab Sahib. How often am I to stress this? You *must* have a son. It is my unfortunate fate not to be able to give you one who will live and grow up to satisfy your ambitions and be a comfort in your old age.'

The Nawab Sahib tried to interrupt her by a gentle pressure of his hand on her lap, but, pushing it away, her eyes blazing, she continued: '*Ai hai!* Do not interrupt me; you always take it as a joke. Listen to me! Just think of what our home will be like when both our daughters are married and gone. It will be like a desert without an oasis. You must now give serious thought to marrying a second wife.' Here she paused, as if meditating, as she had often done before, on all the different possibilities. She knew it would be difficult to get a girl from their own class, for what parents, especially in these times, would care to give their daughter in marriage as a second wife, whilst the first was still alive? Besides, the Nawab Sahib was already past fifty. But it would be easy to find some pretty girl of a respectable family whose parents desired to improve their social status by such an alliance. Then she continued aloud: 'Let me arrange such a marriage, Nawab Sahib, a wife who will be a younger sister to me. Your son will then be *my* son too.' The Begum spoke earnestly. There was physical pain in her heart at the remembrance of the sons that lay buried in their five tiny graves. She was now forty-five and had borne eleven children, the last infant dying two years ago, after one short glimpse of the world.

The Nawab Sahib was more reconciled to his fate than his Begum, and had no desire to marry a second wife and have a son by her. To divert her thoughts, he said with a lightness he was far from feeling: 'Well, well, we shall discuss that later. There is plenty of time yet,' with a cynical emphasis on the word 'plenty'. 'Let us, for the present, discuss the more pressing question of Mehrunnissa's marriage. Has the *mushata* been here lately?' This was the professional woman who played the most necessary role of go-between when marriages had to be arranged.

'Yes, I wished to speak to you about it, but you began other topics and it slipped my mind; my memory is like a sieve nowadays. The *mushata* called only this morning. She says she can find no words to express all the excellence of Mian Zaki and his family.'

'It is her sacred duty to praise all prospective suitors in such high-flown terms; else, what would become of her profession?'

'And your duty is to remain sceptical!' she rejoined, somewhat annoyed.

'Yes, but this time, for fear of being slain by my Begum's tongue, I have had the whole city scoured for information. This last year I must have made hundreds of enquiries regarding your daughter's suitors,' he said, emphasizing 'hundreds' with relish. 'I am well qualified now to set up a private detective agency,' he laughed light-heartedly, 'but daughters are a responsibility, and beautiful daughters ....' After a long pull at his hookah he continued with a forced sigh: 'I really do not think it worthwhile having a beautiful daughter.' The pride in his voice belied his words. 'And it is worse having a beautiful wife; she enslaves one!' With a fond look at her, he added: 'You enchained me with your eyes with the first glance you designed to cast on me, and I have been in bondage ever since.'

'Yes, only the links were weak and the chain has needed constant repairing!' retorted the Begum Sahiba with a half-indulgent, half-upbraiding look.

The Nawab Sahib's deep affection had never really wavered, but flushed with wine, he had occasionally been unable to resist the languorous grace and enticing glances of the dancing girls who sometimes enlivened the men's parties. He had genuinely repented such lapses afterwards.

'Begum, only a man can understand what temptation means. But all that is ancient history. Cut short your memories.' After a brief, embarrassing pause, he went on: 'But aren't we forgetting Mehrunnissa again? Mian Zaki, from all accounts, is most suitable. He is said to have an excellent character, and is a simple, good-natured boy with no overindulgence in learning.'

'Do you think I commend ignorance? Shame on you, Nawab Sahib!'

'Pray, swallow your anger, Begum, and let me proceed with the goodly qualities with which reports credit your prospective son-in-law. You have already heard of his family. Their annual income is over a lakh of rupees. Your daughter will want for nothing.'

'Allah be praised for that! Daughters are dumb creatures placed in our hands as sacred trusts. We can only do our best; the rest lies in His hands.' She raised her arms and her eyes fervently in a gesture of pleading and surrender to the Giver of all things.

'Now that Mian Zaki has so much that could please Mehrunnissa, I hope you will make some allowances for him. Your standard of good looks is much too high for us poor men.' He cast a deprecating look at himself. 'Besides, if he were really handsome it might create rivalry between husband and wife!' he said in a bantering tone.

The Begum gave a short, pleased laugh but which, nevertheless, contained a touch of vexation. The Nawab Sahib's categorizing of himself was by no means correct. He was certainly much darker in complexion than herself and he had passable features, the nose a little too big, the chin somewhat weak. The brown-black eyes, however, were full of good humour, although if roused to anger they turned into piercing darts whilst his whole being burst like thunder. Fortunately, those occasions were rare. He was slim and of medium height and moved with an easy elegance.

'I ought to go now,' said Nawab Sahib. 'The hour of afternoon prayer approaches, and I have kept the estate manager waiting.' Her husband's conversion to prayers was recent and it gave the Begum a thrill of satisfaction. Though always a devout Muslim, as far as convictions were concerned, the Nawab Sahib had seldom before put Islamic precepts into practice.

She remembered the days in his youth, when he had allowed himself to get intoxicated. When she had rebuked him for disobeying the rules of Islam, he had laughed away her scruples by quoting the poet Ghalib: *Can faith be drowned in a bowl of wine?*

Nawab Safdar Yar Jung now rose and, straightening himself, put on his slippers. 'Begum, with your permission, I shall presently return,' he said with exaggerated politeness, implying he wanted to depart and, without waiting for a reply, he walked towards the men's apartments with a leisured dignified tread.

3

A gentle breeze floated through the courtyard, and softly touching the jasmine and the rose, carried a faint perfume, infinitely refreshing, to the zenana reception room, which was open to the courtyard.

'Allah be praised!' said Zubaida Begum with a sigh of relief. 'This is like a breath of heaven after a scorching day of hell!'

Everyone had worked feverishly throughout the day, cleaning and preparing for the following morning when Mian Zaki and his family would come for his engagement ceremony with Mehrunnissa.

Mehrunnissa's modesty, having prevented her from taking part in the preparations, restricted her mostly to her room. The Begum Sahiba had been busy since dawn, supervising her household servants. Now, as she sat discussing the arrangements with Unnie, her ever-eager advisor in all things, she called: 'Gulab, where is Zohra Bibi? Ask her to come here if she is free.' Then turning to Unnie, 'I have hardly seen her face today. How shall I bear the parting, Unnie, if I do not see the girls even for a few hours each day? Allah knows how my heart fills with strange misgivings when Zohra is even a few minutes late from school; my mind turns on fears of all kinds of mishaps. It is a great relief to know that you are always with her—no one like my old and trusted Unnie.'

'May you live long, Begum, and may you be fated to clasp your grandchildren and great-grandchildren in your arms. Although my legs are dangling in the grave, I shall consider myself fortunate if it be written in my kismet to see my children married. I shall then die contented.' She always called Zohra and Mehrunnissa 'my children'.

'Unnie, you shall live to be at least a hundred years!' said the Begum with a smile, and folding a paan, handed it to her. Zubaida Begum always carried her silver filigree paan box with her and, as was the custom with Hyderabadi women, throughout the day folded a paan for whoever may be with her. Unnie bowed and, touching her hand to her forehead with a salaam of thanks, took the paan and put it into her mouth. The Begum continued: 'You are young yet. What do years matter? In spite of your being my Unnie, you look younger than I do.'

'*Owi*, Begum Sahiba, you mock me,' Unnie smiled, trying to protest, but highly gratified. 'Besides, what does a person like me want with a long life? My work is done. I must now look forward to my grave.'

'Oh, no, Unnie, you have not finished your work and we need you. I do not know how I should manage anything without you. You have been like a mother to me.'

Unnie grinned with satisfaction. She knew the Begum was greatly attached to her. Unnie had nurtured her during her childhood and had been her advisor and friend in the early days of her marriage, having come to her husband's household as a young and inexperienced bride of sixteen.

If, at times, Zubaida Begum resented Unnie's high-handed manner, she knew she would just as often disregard it. For where could one find another like her? Such devotion came only through lifelong service in the family.

Gulab went in search of Zohra and found her having her Arabic lesson. As was the custom, she was seated on one side of a curtain and the Arabic master on the other. When Nawab Safdar Yar Jung had met this young *moulvi*, a learned Arabic scholar in straitened circumstances, he had, with the impulsiveness of his nature, employed him at once without considering the propriety of this situation. The horrified Begum had reluctantly agreed only on condition that Zohra's lesson be conducted in purdah. Zohra was chanting the Koran in her fresh youthful voice, full of sweetness. As it reached Gulab's ears, she at once covered her head with her dupatta and, for a while, waited in silence. Words from the Koran instinctively commanded respect. Finding Zohra thus absorbed, she soon slipped away, and whilst passing through the corridor, glanced through the latticed window into the men's quarters.

She caught the eye of Mohammed, who was approaching.

'You'll live to be a hundred and twenty-five years old, Gulab!'

'Why?'

'I was just coming round to you with a message from my master, the Nawab Sahib, to Begum Sahiba.'

'Well then, out with it, you idiot!'

'Yes, Gulab, I'm an idiot. That's why you maids can twist me round your little fingers,' said the soft-hearted, round-faced Mohammed, rather crestfallen.

Gulab was pleased by the flattery but, in order to conceal her pleasure, she said, '*Arrey*, Mohammed, when will you learn not to take everything so seriously?'

Mohammed, raising his short, fat body on his toes, whispered near the window in a sentimental voice: '*Arrey*, Gulab, but where have you maids been today? Not one of you looked out of the window even once. It was as bad as if you were all dead and gone. I was feeling desolate. It's good to see your face again.'

'You liar, it's never good to see my face!' she said fiercely. 'Don't I know who you think it's good to see?' Gulab's attractive face was disfigured by marks as a result of smallpox contracted when she was only seven years old.

Mohammed looked sheepish. '*Arrey*, I want you both to come out tonight for a short while, behind the wall—the usual place. We'll then chit-chat a little. My heart is bursting with topics.'

'Bursting it is .... but with what? I'll send Champa alone; it would be better without me,' retorted Gulab, with a pang of jealousy.

'No, no, Gulab, my parrot, what could we do without you? You keep us lively.'

'My parrot, my parrot, indeed! And what's Champa pray?' she asked defiantly.

'*Arrey*, Gulab, you know, she's only a myna—a timid bird.' But even as he said this, his voice became tender, causing Gulab deeper torment.

'Very well then,' she said with mock-politeness, 'if that be your pleasure, but if we don't come, you may be sure we were unable to evade Unnie's surveillance. That old witch! She doesn't even die.'

'*Arrey*, put magic in her eyes!' exclaimed Mohammed bitterly, wishing he really had some magic for the meddlesome Unnie.

'Ah, but the message, Gulab,' he suddenly cried, 'I can't remember now what it was,' and he started scratching his head. 'That's why they say women are like devils, they make us forget everything!'

'You fool! You said it was a message from the Nawab Sahib.' Gulab's pockmarked face beamed with a sense of her own mental superiority.

'Yes, yes, I remember. How stupid of me! What else could it be from my master but that five or six people are staying to dinner.'

'What sort of people?' asked Gulab, her curiosity aroused.

'*Arrey*, you know, poets and artists and suchlike. The usual sort of people who visit the Nawab Sahib.'

Gulab suddenly recollected her own errand and, with a guilty exclamation, ran into the zenana quarters.

'Begum Sahiba, Zohra Bibi is having her Arabic lesson and so I did not disturb her,' she said demurely.

The Begum Sahiba nodded, but before she could say anything, Unnie intervened. 'But where have you been loitering? Surely it did not take you all this time to find that out?'

'*Owi*, Zohra Bibi was chanting verses from the Koran in such a melodious voice,' said Gulab, cunningly, putting on an enchanting smile, 'that Allah knows I could not wrench myself away.' This served the double purpose of impressing the Begum Sahiba with her love for the Koran as well as her appreciation of Zohra's voice. But Unnie was less easy to deceive.

'Don't I know that whenever you go beyond the zenana you stop to peer out at the menfolk? You are a shameless girl, Gulab, you ought to be thrashed!'

'*Owi*, Unnie, Allah forbid that I should peep out of windows,' Gulab retorted feeling resentful. 'But as I was coming, Mohammed called to me to say the Nawab Sahib is expecting five or six guests to dinner.'

This immediately diverted Unnie's thoughts, and Zubaida Begum, looking worried, said: 'I wish today at least the Nawab Sahib had not pressed them to stay. Tomorrow is Mehrunnissa's engagement ceremony and there is still so much to do before that.'

'I do not think it takes any pressing to make them stay.' Unnie gave a very superior sniff and then, in an altered tone, observed: 'They talk sweetly and take advantage of our master's generous nature.' In this, Unnie revealed the snobbery of servants working in aristocratic households. Had all the guests been nawabs, no trouble would have been too great for her, but for impoverished people, she had only contempt.

'Unnie, their kismet is not very favourable. Let twenty such come; they will all be welcome,' said the Begum Sahiba, at once realizing her obligations; 'I first thought perhaps it would be difficult, after the long day's work and the engagement morning ahead, but what does it matter, Unnie? Ask the cook to prepare extra fowls for the guests.'

Another message arrived from the Nawab Sahib, announcing a *mushaera*, a poetry-reading session, after dinner in the men's quarters and, if the ladies wished to attend, they should make purdah arrangements. The Begum Sahiba now was really angry. Her husband had no idea of what it entailed to organize her daughter's engagement ceremony. She felt weary after the day's exertions but the Nawab Sahib was the master of the house and she had to comply with his wishes. The girls, however, were enthusiastic in spite of the onerous duties of the following day.

Dinner was late that night, and although they were tired, most of them hastened into the adjoining room, divided by a curtain, and without switching on the lights, looked out from behind it. There was a gathering of forty or fifty people all seated in a semi-circle on a carpet spread on the floor. The Nawab Sahib had sent word to all his friends and neighbours, inviting them to the *mushaera*. Amongst the guests were seated the poets, some with more talent than others. But all showed off their emotions and ardour with their every gesture, all waiting for an opportunity to be noticed and all aspiring for recognition.

In the centre of the room stood a lighted candle on a silver salver, which would be placed in front of the poet whose turn it was to recite his poetry.

The recital started on a subdued note; but gradually both the audience and the poets began to warm up. Meanwhile, paan and cups of hot tea were served and hookahs were passed round. In the excitement of the *mushaera* nobody thought of sleep. Zubaida Begum had not the heart to force even Mehrunnissa to retire, keen as she was to have her look her best for the morning's rituals.

It was past midnight. The candle was placed before a youthful poet, who rose to his feet, and cast a swift glance around, to gauge the mood of the audience. He quickly noted and took into account the movement behind the curtain. Slowly, he started to recite. He was a member of the old, impoverished aristocracy and had taken for his pseudonym, Majnoon—'the touched'. It was also an epithet of the celebrated lover of Laila, which made it evident the young man was very aware of his handsome looks.

There was nudging and giggling in the zenana as all eyes turned on him.
'Allah, what a fine bearing!' whispered Mehrunnissa.

'Yes,' said Rashedah, her cousin, 'and what a face and head! Aloof and romantic-looking. But, Apa Jan, it doesn't become a betrothed bride to be effusive about a stranger. You should go and rest and look like a smiling bud in the morning. What do you say, Zohra?'

'You are becoming poetic in the company of poets! But, hush, let's listen.'

'Zohra is so rapt, because he has a passionate voice,' Mehrunnissa grimaced.

But Zohra did not heed them. She was listening intently.

He recited ghazals—lyrical couplets—with a devotional fervour. His dramatic gestures were profuse. The audience worked themselves up into a frenzy of loud ecstatic approval. Majnoon, beaming, responded with low salutations, touching his hand to his forehead.

'Encore! Encore!' was the refrain all around.

'By Allah, Majnoon Sahib is divine! The acclaimed poet, Hafiz himself, could not have done much better!' exclaimed the Nawab Sahib, delighted with the new discovery. Rising and stepping forward, he warmly embraced him.

Zohra was deeply stirred. Her eyes shone with a heightened sense of excitement. She gazed through the chink in the curtain, wishing she could talk to the composer of such verses. His colourful personality had indeed lent charm to his muse. She thought of her own writings, which she had never shown to anyone, except for a few to her father. But if she had the opportunity, she thought she might share them with this poet. She became wistful, thinking how pleasant it would be to meet someone of her own age with whom she could communicate.

The *mushaera* ended in the early hours of the day of Mehrunnissa's engagement and, by the time the last guests had departed, the muezzin's call announced the hour for prayers. A short rest was all they could have before rising on this auspicious morning.

Mehrunnissa's prospective in-laws arrived. The menfolk were received in the men's quarters by Nawab Safdar Yar Jung, while the ladies had travelled in closed cars and carriages with their maidservants who were also anxious to see their young master's future bride. They alighted at the zenana door and Unnie escorted them to the room where the Begum Sahiba received her guests.

After a long wait, Mehrunnissa entered the zenana reception room, supported on one side by her aunt and on the other by a married cousin. She moved slowly, her head bowed, her dupatta drawn over her face. All eyes turned in her direction and the ladies exchanged glances of approval. The would-be-bride's medium height and well-modelled figure appeared perfect to them.

She was led to a small rectangular *takht*, which raised her above the others. The prospective bridegroom's people clustered round her. Her aunt lifted her dupatta, and raised Mehrunnissa's face, supporting it on the palm of her hand; the elders gazed at her in rapturous silence, but the young girls could not contain themselves. They nudged each other and whispered:

'Allah, how beautiful she is!'

'The bride is like a fairy!'

'No, she is like an angel. Allah, I have never seen anybody so lovely!'

'I wish she would open her eyes. I am sure they are enticing,' said Maryam, Zaki's younger sister, enchanted beyond her expectations.

'Ask Mother if she can do something about it.'

Whereupon Maryam edged up to her mother and whispered in her ear, but the mother silenced her with a cold stare and snapped back in a whisper: '*Owi*, Daughter, are you in your senses? If I made such a request, what would they think? They might feel we suspect some defect, which Allah forbid! Besides, how can such a sweet, modest girl open her eyes before us? Borrow some sense and practise patience!'

Her daughter moved away disappointed, for the girls had heard about the beauty of Mehrunnissa's eyes. Anyhow, they continued to make ecstatic observations, some of which were audible to Mehrunnissa. The compliments fell as pleasantly on her ears as raindrops on parched lips.

The engagement jewellery was put on Mehrunnissa; it consisted of a pair of gold anklets set with uncut diamonds and a diamond ring, which was slipped onto the ring finger of the right hand.

Zohra stood in the doorway with her cousin Rashedah, watching the ceremony. As unmarried girls, it was not becoming for them to come forward. Zohra looked very grave as her eyes took in the scene. Would she herself have to undergo similar trials? Her heart trembled at the idea. Besides, how could she stand such close inspection? One needed beauty like her sister's, and the consciousness of it, to go through the ordeal so composedly. How well Mehrunnissa's shell-pink dupatta became her. In her mind she compared her face to a lustrous pearl clinging to its delicate tinted shell. The lack of heavy ornaments, worn only by married women, pleased Zohra's aesthetic sense as she remembered the lines:

> *No ornaments does she need whom Allah hath adorned with the gift of Beauty. Behold the moon's loveliness! Can ornaments enhance her chaste purity?*

As Zohra stood drinking in and meditating on her sister's perfection, she felt far more nervous on her account than did Mehrunnissa herself.

Mehrunnissa's lips were pressed upwards as if in a smile, and they remained fixed. She allowed her face to be lifted and scrutinized without any appreciable effort to lower it, but her eyes remained closed. Zohra, leaning against the door, anxiously twisted and played with the edges of her blue crinkled dupatta, winding and unwinding it round her hands and between her fingers. Her large liquid eyes wore a faraway look, the long eyelashes veiling them partially.

At last, Mehrunnissa was escorted back to her room by Zubaida Begum, and Zohra and Rashedah followed her. She left Mehrunnissa with the girls and hastened back anxious to listen to any comments that might slip from Mian Zaki's family.

Mehrunnissa immediately flung herself on the bed and pressing her heart tightly, exclaimed: '*Ai-hai*, Zohra, Rashedah, get me some water, quick! I am suffocating!' Zohra had expected more restraint from her sister after her sensible behaviour in front of her in-laws to be but, suppressing her irritation, she picked up a fan and started to fan her.

After sipping some water, Mehrunnissa said: '*Owi*, Zohra, how can you ever know what I have been through? And the comments! They were loud enough even for me to hear!' She tried to sound indignant, but the note of pleasure in her voice was not lost on either of her listeners. 'Although my eyes were closed I could feel their gaze burning my face. I felt scorched all over. *Ai-hai*, what an ordeal! I must have looked quite silly. Did either of you see me?' She pretended to await the answer with anxiety. Zohra, for once, felt she should forgive Mehrunnissa her vanity. But before she could form a reply, Rashedah burst forth effusively:

'Apa Jan, you looked so stunning that we stood rooted in the doorway. Alas for my fate, but that I were a boy!' They all laughed.

'*Owi*, Rashedah. What are you saying!' exclaimed Mehrunnissa affectedly.

'I marvelled at your self-control,' said Zohra. 'Not a lip twitched. Not an eyelid fluttered. I would have made a fool of myself, but then, Apa Jan, I have not your beauty.'

'Nonsense, Zohra!' Mehrunnissa contradicted happily, 'as for my self-control, Allah knows how I wished the earth would open and swallow me up. But, Zohra, Rashedah, didn't either of you overhear any remarks those people made?'

'They all clustered round you, but I'm afraid we did not belong to the privileged circle. I was careful not to let even my shadow fall near them lest they should think you were like me!' laughed Zohra, appreciative of her sister's handsomeness. 'But, Apa Jan, tell us what you overheard.'

Mehrunnissa had been leading them up to this, for she was bursting to tell them. She began the recital with a triumphant smile.

When it was finished Zohra suddenly said: 'Rashedah Apa, let us try and slip into the corridor. With a little luck, we might get a glimpse of Apa Jan's betrothed.'

'We'll have to be wary,' said Rashedah, 'for were his people to catch us watching, we should die of shame.'

Mehrunnissa heard this in bashful silence, and for the first time that day, felt an interest in someone other than herself. She wished she too could join them.

Zohra and Rashedah only caught a fleeting sight of Zaki as he entered the carriage, and full of excitement, they rushed back with confusing impressions, which, instead of satisfying Mehrunnissa, only annoyed her.

Zohra, now feeling sorry, said: 'Apa Jan, we were so eager to see everything in so brief a moment that I fear we saw nothing, but I promise to get it out of father, for when he comes in, Abba Jan will relate everything to you faithfully.'

Even after Zaki's people had left, Nawab Safdar Yar Jung was detained in the men's quarters by his friends. Zubaida Begum, anxious to discuss the day's happenings, sought out Zohra and her niece, Rashedah.

'Chachi Amma, first sweeten our mouths with sugar and then we shall tell you what your son-in-law looks like!' exclaimed Rashedah in her lively manner.

'What! Did you girls try to get a look at him?' Her aunt tried to sound disapproving, but the eagerness in her voice gave her away.

'Chachi Amma, if we have acted unwisely, perhaps we had better keep it to ourselves.'

'*Ai-hai*, Rashedah! You devil! You cannot help making a fool of even your old aunt. Out with what you have seen, and I shall not only sweeten your mouth with sugar, but will fill your lap with it. Quick, make haste!'

But when they started their tale, amidst bursts of laughter, they only contradicted each other. One said his eyes were brown, and the other black; one had found them big and the other small; one thought his lips a little too thick and the other fine; one said the nose was too broad and the other that it was ordinary. At last the Begum Sahiba in despair interrupted:

'*Owi*, girls, are you sure you have been looking at the same man? You sound as if you had seen twenty different men. What intoxication thus confused your minds?'

The Nawab Sahib hastened in as soon as he could, and was more helpful in satisfying the Begum Sahiba's curiosity. Mehrunnissa afterwards learnt of it all through Zohra, for marriages and future husbands were not subjects to be discussed between parents and daughters. Convention held such talks to be immodest.

This suited the parents, for they could then select partners for their children, with maturer judgement, and without the emotional counsels of youth.

As the wedding day drew near, Mehrunnissa looked happy and excited. The idea of marriage caused her no anxiety, only curiosity.

4

Nawab Shaukat Jung Bahadur's daughter, Safia, sat facing her mother as she eagerly announced that she had at last found a suitable bride for her brother Bashir. Safia had been at the Mahbubia School at the same time as Zohra but, as she was several years older, the girls had had little to do with each other. Nevertheless, Safia had impulsively marked her out as a desirable bride for her younger brother Hamid.

Hamid, however, was still in England and although he had procured a good degree from Cambridge University, he did not wish to return to India yet. In spite of his mother's pleading letters, he resolved to stay on for another few years. He had made many friends and the free and easy life that he was able to lead suited him well.

Safia had, by chance, one day met Rashedah and, whilst exchanging family news, had learnt that Zohra's parents were anxious to get her married soon which meant there was no chance of waiting until Hamid's return.

Safia's thoughts had turned to her older brother, Bashir, for being the only sister, she had been entrusted with finding him a suitable bride.

Bashir had returned from abroad after obtaining a doctorate in physics at Cambridge University. Until now he had refused to consider marriage arranged by others. Four years in Hyderabad had, however, modified his attitude. There seemed no other way of marrying unless he looked for a bride outside his own background. He was tired of his bachelor life.

In spite of a considerable difference in their ages, Safia decided Zohra would be an appropriate match for Bashir. As she now put this proposition to her mother, there was an air of expectation amongst the women of the zenana. Masuma Begum listened intently to her daughter and, after some discussion, Safia gained her consent to try and gauge her brother's mood. She lost no time and walked straight into his study, and perched herself on the arm of a chair close to him. Bashir was at his desk, absorbed in work and seemed not even to notice her.

'Bhai Jan!' she called.

'Hmm,' he grunted, without looking up.

'Bhai Jan,' she said loudly, 'Bhai Jan, listen to me!'

'What?' he asked, his eyes still on the book. He was writing notes.

'*Ai-hai*! Turn to me. Surely I can't talk to the wooden back of your chair! I've come on an urgent errand. I've come to tell you of a bride I've found for you!' she almost shouted as if he were deaf. 'Only, I think she is much too good for you!' she snapped, piqued at his continued inattention. Ordinarily, she would not have dared to disturb him at his work.

After a moment longer he carefully pushed the book to one side and, putting down his pencil, turned himself around in the chair and said:

'That doesn't sound promising. I haven't the least desire to marry anybody too good. Angels and houris, those beauties that dwell in Paradise, are for God's chosen few. I want to marry an ordinary, healthy human creature of flesh and blood.' Whilst apparently still absorbed in his work, Bashir had taken in every word Safia had said. He had the power of simultaneously giving attention to two things when necessary. Safia could never get used to this and gazed at him wide-eyed. He merely continued:

'You have to remember, Safia, that I'll be making my home on earth and not up in the skies; your houris wouldn't even get acclimatized here.' He spoke in a flat, unemotional voice, focusing his keen eyes on her.

Safia always felt belittled when Bashir talked like that. Still seated awkwardly on the arm of the chair, she started slowly swinging her right leg.

'Why don't you sit down properly?' he asked, looking disapprovingly at her.

'No, I'm all right,' she said stubbornly, determined not to let him intimidate her at least today. 'Besides, I have other work to do, and shall be off presently.' Her voice sounded annoyed but he took no notice of it.

'Well, who is this girl you were talking about, anyway?' he asked.

Perceiving his curiosity aroused, she slipped her angular body into the armchair and began a glowing account of Zohra. She pressed him to consider the prospect so that a proposal could be sent immediately for fear of losing her to some other young man.

At last he said: 'May I at least first see a photograph of—according to you—my prospective wife?'

Safia had known Rashedah since their school days and, through her, had obtained a photograph of Zohra. It was an ordinary snapshot taken by her cousin Sagheer. Safia placed it on the page of the open book on his desk. With a slight inclination of his head he scrutinized the picture, almost as if he were continuing to read the book. Safia, watching him eagerly, could divine nothing from his immobile face. When he spoke, his voice was measured but listening for the slightest reaction, Safia could detect a hint of interest in it.

'She looks all right,' he said.

Safia, gathering courage, asked with affected nonchalance: 'Bhai Jan, through all your long travels across many lands, have you never felt romantic or sentimental about anyone?' She was stretching her neck forward like a crane.

'I consider romanticism and sentimentality to be the products of a diseased mind,' said Bashir in his smooth voice, 'and women, being the weaker sex, succumb to them more easily. Reality is above such foolish imaginings.'

'So, you've never been foolish?' asked Safia, stretching forward still farther.

He lit a cigarette and after a puff or two, quickly said: 'Since you are so anxious to know, I confess I did make a fool of myself once. But that was a long time back, when I was fresh in England and didn't know the ways of women. After that, I learnt my lesson. I hope I am immune to such temptations.' He sounded superior, like an experienced man speaking of some childish prank. He resumed his cigarette.

To Safia, all this sounded strangely remote from life, but having no wish even to attempt an argument with her brother, she withdrew.

When Safia had left the room, Bashir turned thoughtfully to the photograph. He was struck by Zohra's distinguished air. The eyes seemed to draw him to her, for they looked straight at him and yet had a shy, withdrawn expression he found intriguing. The lips smiled at him, but with restraint. Against his will, he was moved by it.

Bashir, having little faith in Safia's judgement, made his own enquiries from other sources regarding Zohra and her family, and as he was satisfied, a proposal of marriage was sent.

Months passed and there was no reply. Bashir, having once made up his mind, was exasperated by this delay.

One day he asked Safia: 'What does all this mean?' His level voice betrayed an undercurrent of anxiety.

'After all, they must also find out things about you,' replied Safia secretly delighted to see her brother show some ordinary human emotions.

'But what is there to find out about me that takes so long? There must be some skeletons of which even I am not aware.' His pride was hurt.

'Bhai Jan, these are the ways of the nawabs. They will take their own time. You know the old saying: *The soles of one's shoes must wear off in going to and fro before a favourable reply can be expected.*'

'But what if for some reason the answer is not favourable?' asked Bashir, trying to sound detached.

But Safia, now daring to look at him with open amusement, said: 'Delay in such matters, they say, is always a good omen.'

5

*M*eanwhile, in the household of Nawab Safdar Yar Jung, the excitement seemed tangible. Ever since Mehrunnissa's marriage three years before, Zubaida Begum's mind had been set on finding a suitable bridegroom for Zohra. She discussed the subject incessantly with her husband the Nawab Sahib, with Unnie, and with friends and relatives. It had become a complete preoccupation with her. Then one day, during the fasting month of Ramazan, the *mushata* arrived carrying a proposal of marriage from the distinguished family of Nawab Shaukat Jung Bahadur. Zubaida Begum was in her element. Now it only remained to gain Zohra's consent.

One evening, Rashedah in her usual breezy manner, walked into Zohra's room. Seeing her sketching, she exclaimed:

'*Owi*, Zohra, whoever heard of anyone painting whilst fasting? You delight in buying yourself headaches.'

'Anyhow, headaches are better than heartaches!' Zohra laughed back.

'Much you know about hearts,' exclaimed Rashedah, 'shutting yourself up in a lifeless world of your own!' Rashedah was one of those fortunate ones whose parents' mature judgement and her own youthful inclinations had unconsciously worked in unison. For her parents had recently got her engaged to her cousin Sagheer. They had played together as children, and fallen in love as they grew up, not aware, in the beginning at least, that the parents on both sides had planned their union. But after their formal engagement, Rashedah could only meet him secretly, with Zohra's help. Therefore the two girls had been drawn closer together. Looking at Rashedah with amusement, Zohra asked: 'Rashedah Apa, what good fortune brought you here today? Or have you arranged a rendezvous with Sagheer under my benign protection?'

'*Ai, towba*, who wants to meet him! I came to see you.' Rashedah shook her head coquettishly, vainly trying to hide her pleasure at the very idea. 'But, Zohra, what about your bridegroom? Do his qualities please the fastidious bride?'

Zohra gave a start; her eyes looked like those of a deer suddenly awakened from slumber by the noise of gunshot.

'What?' she asked instantly, as if she had hardly understood Rashedah.

'Why, Zohra, I had no idea you would be so startled. You know that discussions have been going on for months.' Rashedah was apologetic for so abruptly introducing the subject. 'Don't look as if you were doomed to something dreadful, Allah forbid. He has all the qualities that will make you happy. Both Chachi Amma and Chacha Jan are agreed that there can be no better match for you.' As Zohra listened dazed, Rashedah continued: 'They are anxious to arrange the engagement before the festival of Eid at the end of this month of fasting, so that the wedding may be celebrated before the start of the mourning period of Moharram. But they want you to agree.'

Zohra's face was now drained of all colour.

'Rashedah Apa, why do they want to rush me so? I have no wish to get married. Why can't my parents wait?' There was a note of rebellious exasperation.

'Shame on you, Zohra! Do you think it gives parents pleasure to deprive their homes of children? You, an intelligent girl, shouldn't make it more difficult for them.'

'But, Rashedah Apa, one has one's whole lifetime for marriage; a year or two should make no difference to Ammi Jan and Abba Jan. It will make all the difference to me. I am in no mood to marry. At present, painting interests me far more than the idea of marriage, and Miss Woods is a wonderful art teacher. Lately, we have done a few landscapes together and I find myself enjoying our classes more each day.'

'Oh, I see. Now Miss Woods is the angel. If she is really so wonderful, so lovable, so gifted, why couldn't she find a husband for herself in England instead of coming all the way to India and putting ideas into our innocent girls' heads?' Rashedah was half earnest, half bantering.

'You can think of nothing but marriage!' retorted Zohra indignantly. 'She did not marry because she didn't want to.'

'Oh yes, I can easily imagine Miss Woods with young swains clustering round her, all sighing for her favours,' said Rashedah, more to tease Zohra than out of malice, for she too liked Miss Woods. But the picture of that now-greying lady, surrounded by supplicants, tickled her imagination.

'Don't be silly, Rashedah Apa. But I am not joking. She could have married several times if she had wanted to. Only there was nobody to her liking, and why marry unless one is really fond of someone?' Realizing she had unwittingly exposed her feelings in the matter, Zohra hurried on: 'Besides, there was a man with whom she was in love, but he was killed in the Great War. She never found anyone else for whom she could care in the same way.'

There was something in this that at once touched Rashedah's heart, and she became serious, and was sorry she had spoken so lightly of Miss Woods' imaginary suitors. After a brief pause, she pursued the subject:

'I think it's folly to try to put off one's kismet. And after all, marriage is the fate of all girls. Why delay it unnecessarily?' She was now echoing her aunt's sentiments. 'His name is Mohammed Bashir and he is the son of Nawab Shaukat Jung Bahadur. A more eminent family would be hard to find. You should be happy, Zohra, that your parents are not marrying you to some old-fashioned nawab, but a man after your own heart; a man who was a brilliant student in England, and now has a bright future before him. You, who have filled your mind with western ideas, should be grateful for this. He will surely not interfere with your hobbies even after marriage.'

'After marriage!' repeated Zohra in a tone in which there was despair and foreboding. Regaining her self-control, she said in almost a whisper: 'Rashedah Apa, hasn't he a younger brother still studying in England?' She recalled a brief conversation at school with Safia, the sister. Then hesitantly, as if afraid of her own boldness, 'I might marry that brother when he comes back.' She gave a little nervous laugh. 'That would, at least, delay the fateful day.'

'Chachi Amma considers unnecessary delays foolish. Besides, you know she complains of palpitations, and although doctors think it's only a case of nerves, she constantly worries about your future, which makes her worse. Allah forbid, but if anything were to happen to her, she says her soul would find no peace, knowing you were still unmarried and alone in the house.'

Zohra's face became grave as she listened to Rashedah. She knew how her mother had suffered through the loss of children, and was anxious to see her daughters' lives fulfilled with children of their own. But marriage, Zohra knew, would mean the end of her freedom. The vision of a marriage of one's choice tantalized her, although there seemed to be no possibility of it.

After a while, Rashedah again started to press her point, for after all she was her aunt's emissary.

'Besides, Zohra, you know your parents have their hearts set on performing the holy pilgrimage. But how can they go to Mecca unless you are married and settled in your own home? The thought lies heavy on their hearts.'

Zohra's face was ashen as she bowed her head, holding it between her hands. It seemed to whirl round, and her hopes with it. If only her parents could understand what ideals she had to sacrifice! But in the present scheme of things she knew they had every right to get her married as they thought fit. It was only out of consideration for her feelings that they were trying to gain her consent.

Rashedah, not knowing how to handle the situation, sat silently on the settee.

After a long pause, Zohra raised her head, and leaning back in the chair, said in a voice drained of all emotion: 'Rashedah Apa, please tell Abba Jan and Ammi Jan that I'll do whatever they wish me to do. I cannot let them worry over me.'

'Zohra, don't talk as if—Allah forbid—it were a mournful occasion,' was all Rashedah could say.

Zohra said wearily: 'Convention is a strange phenomenon. Here you are, in love with Brother Sagheer, standing, as it were, on one foot to marry him, and you are compelled to wait, whilst I am forced to marry a stranger in this wild haste. Why does the world treat us so unfairly?'

'Do you think I am in the world's confidence?' asked Rashedah, attempting a lighter vein again. It was getting late. Through the windows they could see the sun setting, and Zohra, watching it go down, thought, 'My hopes are setting with it.' To her companion she said: 'Let's prepare for evening prayer.'

But once alone, she lost all self-control and wept bitterly and long. The loud boom of a cannon marking the time to break the fast, brought her back to herself. Hurriedly she made her ablutions, whilst the sonorous voice of the muezzin sounded from a nearby mosque, summoning the pious to prayer.

'*Allah-ho-Akbar; la-Illaha-il-Allah!*'

'Allah is great; there is no God but Allah!'

These simple words, at least momentarily, fell soothingly on Zohra's ears, helping her in her effort to control her agitation ... Yes, Allah was great, and as there was no other God, she would place her trust in Him.

Zohra broke her fast with a pinch of salt. Then, after a few minutes of prayer, during which she tried to find some special consolation, she quietly went into the dining room where her mother and Rashedah were already seated with a feast spread before them, for since dawn they had touched neither food nor drink. In the democratic fashion that a few Muslim homes still preserved, Unnie and the other women-servants and maids joined them in the repast.

II

The month of fasting was at an end and the next day would herald the festival of Eid. Bashir's patience had at last been rewarded. The marriage proposal was accepted and the families rejoiced at such a suitable and honourable match.

In the zenana courtyard, the women of the household were gathered to catch a glimpse of the Eid moon. Eager eyes were scanning the sky.

'*Arrey*, look, there it is,' exclaimed Gulab, dancing around excitedly, pointing to the fine crescent.

It reminded one of the eyes of a newborn baby that first blink through narrow slits, catching a glimpse of the unknown world.

Seeing the excitement in the courtyard, the Begum Sahiba also stepped out; but immediately her eyes glimpsed the crescent, she covered her face with both hands; 'Zohra, where is Zohra Bibi?'

'Here, Ammi Jan,' said Zohra moving closer to her mother who, uncovering her face, rested her eyes on her daughter; it was auspicious to see somebody one loved dearly, immediately after viewing the moon. But, tonight, this simple ceremony had a greater significance, for it was the last Ramazan Eid moon, before Zohra's marriage. Both mother and daughter were sadly aware of this. The Nawab Sahib came in hastily, saying, 'I came to view the new moon from here.'

The Begum Sahiba pointed out the now brightening crescent, looking surer of itself. He took a quick glance, then turned his gaze on Zohra. She, moving close, bowed her head before him to receive his blessings. The Nawab Sahib, placing his right hand upon her head, said with unaccustomed gravity: 'Allah be praised for this Eid moon. Zohra, look, there is her faithful companion, your name-star.' He was pointing to Venus. 'Allah willing, the star of your kismet will shine as brightly throughout your life!'

The Nawab Sahib, after gazing at the newborn moon, turned to his daughter: 'You remember the poet Iqbal addressing his Muslim brethren, says, *The sword of the Crescent is our National Symbol?*'

'Yes,' continued the Nawab Sahib musingly, 'the crescent is our guardian-sword, our guiding light. This somehow reminds me of Majnoon, the poet from Delhi, who recited his poems on the eve of Mehrunnissa's engagement. Zohra, do you remember him?'

'Yes, Abba Jan,' was the quiet reply, respect for her parents not permitting her to confess to the keen interest she had felt in that romantic figure.'

'I took him to meet our prime minister this afternoon. You know he is a great lover of the arts, especially poetry, and composes verses himself. He was charmed both with Majnoon and his poetry, and has invited him again. He even suggested presenting him to His Highness, the Nizam, and, Allah willing, something may come of it,' said the Nawab Sahib with obvious delight at the prospect. Their talk came to an abrupt end for Unnie and the other maidservants, who had stepped back until the parental greetings were over, now came forward to offer obeisance.

Felicitations were exchanged all round.

To Unnie, Zohra offered salutations first. Both her age and her status as her own mother's nurse demanded this courtesy. Unnie poured blessings onto Zohra: 'May you live long, Chhoti Bibi, and may Allah make your kismet forever auspicious, and may you see the next Eid moon in your bridegroom's home, doubly happy!' Zohra remained silent. All the other women-servants spoke in the same vein for there was much excitement over Zohra's forthcoming marriage. Zohra listened wearily and slipped away as soon as her father left. Bolting the door of her study from inside, she sat brooding.

'These women!' she thought, as tears crept into her eyes. Why did they go on talking endlessly about her approaching marriage? They harped on the subject as if they had composed some tune especially pleasing to her ear, with which they tried to pass away her tedious hours. If only they knew how it made her head ache. She rose listlessly and stood at the window, dreamily gazing at the star-spangled sky. Suddenly the romantic face of Majnoon presented itself vividly to her mind, and she wondered if her future husband was anything like him. With a sensation of sudden shock, Zohra recognized that the thought of Majnoon had awakened in her a disturbing feeling.

That night, sleep fled from Zohra as if frightened by her restless spirit. In the middle of the night, she sat up, startled and wide-eyed, as if she had suddenly discovered a desert-oasis. Why should she have to go through with this marriage? She could run away. She might meet a young man—yes, like Majnoon, a poet—and fall in love with him. But of course she knew that option was not open to her.

Perhaps she could implore her parents to let her go and work with Mahatma Gandhi. Recently she had developed a longing to be part of Gandhiji's movement of non-violence and she understood the need for the youth of India to fight alongside him. She knew that hundreds of thousands of people flocked to see him, to hear and touch him wherever he took his message of ahimsa. And was he not bringing emancipation to the women of India, even causing a number of Muslim women to shed purdah in the zeal to work for the national cause? But reality in Hyderabad was very different. Politics was not a topic easily discussed in this house, and she knew she must rid herself of patriotic desires.

More realistically, perhaps, her parents would consider letting her study at Shantiniketan, the 'Abode of Peace'. She would develop her art and learn to write poetry at this school in Bengal, started by India's foremost revolutionary poet, Rabindranath Tagore. After all, it was the centre of the revival of national culture, and Tagore was not only a great poet but also a man of true understanding and high idealism. Zohra had only been ten years old when Rabindranath Tagore won the Nobel Prize for Literature, but she still remembered the pride and excitement that had run through the house. And only recently he had surrendered the Knighthood, the British had bestowed on him, as a protest against the massacre at Jallianwalla Bagh in Amritsar. Zohra longed to be a useful part of the events that were stirring her country. Marriage and children should come later. But even as these thoughts sped through her mind, she came face to face with the inevitable truth. She lived in a world where freedom of choice was not available to women. Any deviation from the accepted norm would deeply wound her parents. They might survive the shock but they would never be able to lift up their heads again for shame and sorrow. She had agreed to this marriage for her mother's sake. She could not, in the end, destroy her family. Life must take its destined course.

The next morning, they were all bathed and dressed early. Nawab Safdar Yar Jung went to the mosque for Eid prayer. The Nizam himself was present and all the nobility were also gathered there, and a true festive spirit prevailed.

After the Nawab Sahib's return, a stream of visitors came to offer Eid felicitations. Zubaida Begum was kept busy folding paan and sending them out to the guests on silver platters together with perfume shakers of silver filigree full of rose water, musk, and jasmine.

It was one of the Nawab Sahib's wishes that the Begum herself should prepare paan, as none other, he said, knew the art so well. Pleased with the compliment, she humoured him.

In the zenana, Mehrunnissa arrived with her two sons, Shahid and Anwar. They were dressed in clothes sent to them by their *Nani*—their maternal grandmother—in celebration of Eid. They wore knee-length coats of red brocade, with tight-fitting white trousers. Their heads were covered with gold-embroidered fez-caps, and their feet with pointed slippers. Both the boys looked charming, and Zohra's distracted mind found relief in playing with her little nephews.

Mehrunnissa's beauty was somewhat dulled by three years of married life. Her once voluptuous body was now too ample for a young woman of her age. Her features betrayed a certain heaviness. Moreover, she was expecting a third child.

Tired after the morning's excitement, she was resting in the afternoon in her sister's room. As Zohra played with the younger child, Mehrunnissa watched them, with her ego well satisfied; then she said, affectedly: '*Ai-hai*, Zohra, unmarried girls don't realize how fortunate they have been until they are married. There is no freedom in marriage. One is weighed down by responsibilities. You will remember my words when you're married!'

Zohra well knew that it was only a pose Mehrunnissa assumed in order to make herself sound more interesting. Nevertheless it was true that she was genuinely tired of child-bearing. But that was partly compensated for by the fact that it gave her a sense of importance, for which she craved as an alcoholic does for drink, and she always exploited her condition to the greatest advantage. Apart from bearing them, however, she had little to do with her children. Her mother-in-law took upon herself the responsibility of looking after them with the help of servants and nurses. Mehrunnissa was very happy in her married life. She was always anxious to go back to her husband, and no amount of persuasion would induce her to stay for any length of time at her parents' home. She often made excuses about her mother-in-law being offended if she stayed away too long. But this could not deceive even her mother, anxious though she was to believe all that her daughter told her.

Mehrunnissa's husband was infatuated with her and offered her wholehearted devotion, to which she responded with all her passionate nature.

Watching her sister reclining on the divan, Zohra felt an urge to ask her more about her married life. But deep apprehension flooded her mind and drowned the words before they were even formed.

6

Nawab Safdar Yar Jung's household was in turmoil in preparation for Zohra's marriage. The house was being redecorated and everything had been thrown into disorder, whilst shopkeepers began to send their different wares for selection. The place was besieged by jewellers with beautiful ornaments in both Hyderabadi and western designs; cloth vendors with their exquisite silks; Banarasi merchants with saris in transparent gold and silver sheens. Zubaida Begum bought saris for presentation to Mehrunnissa and Zohra's female cousins and girlfriends. Brocades were bought for the men to be made into sherwanis.

Nor were the servants forgotten, and suitable clothes were bought for every one of them. Men's clothes were given to the tailor but women's garments were sewn at home by poor women from the neighbourhood, specifically called in for this purpose. They responded enthusiastically because, apart from the daily wages, the Begum Sahiba was generous with food and presents. But she was not irresponsibly lavish like her husband. For, now with the wedding approaching rapidly, the Nawab Sahib, desolate by the thought of parting from his beloved daughter, considered nothing too extravagant for her. The shopkeepers, soon sensing this, had been tempting him with all kinds of extraordinary, but useless things. The Nawab Sahib often bought them on an impulse, without even consulting his Begum. Zubaida Begum had been reprimanding him mildly but he, taking it all in good humour, had made no effort to curb his wayward desires. However, one day he walked in to where his Begum was seated, arranging a part of Zohra's multicoloured trousseau into different piles. Sitting down and glancing at these, he said enthusiastically: 'Begum, I have bought something far more beautiful and unique than all this for Zohra.' He gave his moustache a gentle twirl of self-satisfaction.

She then saw that Gulab and Champa had followed him, carrying a glittering basin and ewer. The Begum looked aghast. The maids placed the articles before him and immediately departed.

'But, Nawab Sahib, what is Zohra going to do with all this? She is not marrying a potentate to need such things!' Her voice and face conveyed severe disapproval.

'She is as good as a queen,' said the Nawab Sahib proudly. 'Ordinarily, she can use silver, but this gold is something special!'

'You have no sense of proportion whatsoever,' cried Zubaida Begum in dismay. 'We can't manage all this without incurring debts.'

'Even if there were debts, the sale of a small portion of my estates would help pay them off. I am also ordering the finest car.' The Nawab Sahib, once annoyed, became defiant like a child.

'*Ai-hai*, Nawab Sahib, I have never heard anything so fantastic,' she burst out with some consternation at the thought of having to sell their property.

'I will do as I think fit for my daughter's wedding,' the Nawab Sahib flared up. 'I will no longer tolerate this woman-*raj*!' All of a sudden, feeling impotent in the face of his Begum's logic and thwarted in the desire to give Zohra the very best of everything, he was trembling with rage and pushed away the paan the Begum had made for him. He rose hurriedly to leave the apartment.

As he stepped from the *musnud* to put on his slippers, he knocked down the hookah which fell, spilling cinders and water on the pile of the bride's trousseau close by. But he was in no mood to care. His brown face was red with fury. The Begum Sahiba, watching all this, now turned rigid like a stone idol. Long experience had taught her that such tempestuous outbursts from her husband though very rare and unpredictable, were best left to take their own course. She had gradually trained herself to practise at such times, tremendous self-control. She did not now utter a single word.

The Nawab Sahib, true to his threat, for two days did exactly as he pleased without setting foot in the zenana. But the mood dying its natural death, he started feeling repentant and lonely. On the third day he entered the zenana again. The Begum was seated with a group of women around her to whom she was distributing sewing jobs. As the Nawab Sahib approached, they all rose, respectfully greeting him with salaams and hurriedly disappeared. The Nawab Sahib now offered his salutations to the Begum in the most courteous of manners. She accepted them with great self-restraint.

'With your leave, Begum, I shall sit down,' he said, smiling diffidently. The Begum Sahiba graciously pointed to him his place. The Nawab Sahib ceremoniously placed a bejewelled talisman, made into an armlet, before her. The Begum accepted it equally graciously, and peace was silently restored between them.

But the thought did pass through her mind, with a kind of amused annoyance, that this was hardly the right peace offering after a dispute regarding his extravagant expenses. But the Nawab Sahib could not be dictated to. Zubaida Begum also knew that the reason for her husband's intemperate behaviour was the deep sorrow he felt at the prospect of Zohra's departure. After this, the Nawab Sahib displayed a more reasonable frame of mind regarding bridal gifts.

Several days before the wedding, near relatives, especially the women, came to live in the house to help in the preparations and to give it a festive appearance.

Zohra continued to go to the Mahbubia School until a week before her marriage. The Begum Sahiba wanted her to stop much earlier. 'A *dulhan* must look fresh like a blossom, and Zohra is already looking worn out,' she protested to her husband.

But the Nawab Sahib, who had been watching Zohra, and had heard Rashedah's appeals on her behalf, interceded with his Begum: 'Let her go. What can she do at home? Begum, let her enjoy these last days of girlhood as she wishes. She will have duties enough when she is married.'

The Begum Sahiba, who had had to curb so many of his other wishes, thought it wise to give way to her husband in this.

At school, Zohra found a kind of serenity in the calming company of her friend and confidante, Nalini. It was a refuge from the feverish excitement in the house. '*Shaadi, shaadi, shaadi,*' was on everybody's lips, as if their minds had gone to sleep on all other topics. *Shaadi*—marriage—literally meant happiness. But what a mockery of the word, she thought.

School provided her respite from the ever-recurring refrain. She had implored the girls connected with her family not to mention her forthcoming marriage to her friends and teachers. At least at school she wanted no fuss made about this wedding. The invitations to her school friends were held over until Zohra had departed. A week before the wedding, she left Mahbubia School for the last time and quietly slipped away without telling anybody, but her heart was heavy and she cried silently when once inside the purdah carriage.

Five days before the wedding, the first of the marriage ceremonies began. In the morning, a procession arrived from the bridegroom's home, bearing various perfumed powders—sandalwood and yellow turmeric—which were made into a paste. Zohra, seated on a low settee, with a veil drawn over her face and head bent low, was gently rubbed with the bright yellow paste on the face, hands and feet, by seven young matrons, including Mehrunnissa, both the number and their married state being considered auspicious. Turmeric and sandalwood would soften the skin, healing any blemishes and give the bride a palely glowing complexion. After the ceremony, Zohra was bathed and dressed in yellow clothes, this colour being symbolic of *Manja*, when a bride-to-be went into seclusion wearing only yellow until her wedding day. Her unmarried cousins and close friends, including Rashedah, also arrayed themselves in yellow clothes, presented by Zubaida Begum, for from then on until the wedding day they would be her constant companions. Nalini was also a frequent visitor, going in and out of the house, but she could not stay.

'*Arrey*, Nalini, Zohra is inconsolable without you. You must stay here, until the fair *dulhan* leaves for her bridal home!' exclaimed Rashedah.

'But how *can* I?'

'Why, we'll cook pure vegetarian food for you. I'll make it with my own hands. Are they unclean?' Rashedah displayed her henna-dyed palms.

'Don't be absurd,' Nalini laughed quietly. 'You know I can't displease my elders.'

The house reverberated with the anticipation of the approaching nuptials. The numerous relations added to the lively atmosphere. The older members occupied themselves with sewing, the more talkative ones keeping the party alive with anecdotes and reminiscences from their own bridal days. For at the time of weddings, women's minds invariably wander back into the past, recalling their own or their relations' marriages.

The younger members were in a state of gay abandon. Often they gathered in small groups about the fountain that played in the zenana courtyard. As with the Hindu festival of Holi they mixed rainbow colours in brass pots, and cupping the coloured water in their palms, splashed one another. Maids mingled with mistresses freely joining in the fun. Laughter rang out all day through the house, and the elders looked on with indulgence and contentment. 'Allah, how good it is to be young!' they sighed.

Sometimes, at midday, when the heat was too intense for the girls to venture out into the courtyard, they sat inside and applied henna to their hands and feet. It was cooling. When removed, the red colour left on their palms, pleased their eyes and they vied with one another in getting the deepest tint.

The bangle seller came every day for, other than widows, everyone, old and young alike, must have new sets of bangles on their wrists. The Begum Sahiba, who was presenting these, asked each one to select the design of their choice. They fluttered round the bangle seller as moths around a candle, and while she was there, the maids utterly neglected their work. Zubaida Begum looked on indulgently, but not so Unnie, who was forever seeking excuses to reprimand them. Once, when she saw them all gathered around the open steel trunk in which the bangle seller transported her bangles, she burst forth: '*Owi*, have you all gone mad? There is no one but thinks of her own adornment! As if you all were to be married!' she scolded. 'Not one of you has touched any work this whole week. *Ai-hai*! Has each one of you taken a vow to some blessed saint that she will abstain from work until Chhoti Bibi is married? Do you think that I have ten pairs of hands instead of one—may Allah preserve them—for I earn an honest living by them.'

The maids looked at one another, winked, and made wry faces behind her back. After Unnie had bustled away, with an elaborate show of being in haste, Gulab, looking wretched, exclaimed: 'Does this mean-spirited woman really think she's doing all the work? Indeed she bosses enough for ten, or even twenty. Perhaps that's what she means by saying she does the work of ten!'

'And she took the longest to select her bangles,' added Champa.

'Yes, and we all had to stand by for Her Highness's pleasure! Really, she acts as if she were the super-Begum!' Gulab made a strutting gesture. 'And it was not for herself alone that she selected bangles but for her whole clan of females, daughters and granddaughters, and, Allah alone knows what other relations. They're all coming to stay for the wedding!' Gulab made a wry face. 'Our Begum Sahiba can never say "no" to anyone. Her generosity is limitless.'

'May Allah bless Begum Sahiba! She is born to rule and I would cut off my right hand in her service if need be,' exclaimed another maid, 'but the power Unnie has gained over her is not the rightful share of any human being born in her low state.'

'And it's not as if Unnie were speaking the truth. She waddles around like a fat hen, bossing everybody about and then sits down to eat paan pretending she's overworked,' Gulab's gleaming eyes danced as she gesticulated in imitation of Unnie, sending the others into peals of laughter.

'*Owi*, and we do the hard work,' chimed in Champa who, though timid by nature, was emboldened by the general reaction to Gulab's remarks.

Gulab, with a wink, observed: 'And what about those tiny black pills rolled with pure tobacco, which she chews all the time? She calls them the elixir of life. It's really Satan who's forever driving her.' And Gulab frisked around, as if she were riding a horse, causing more giggles. Conscious that she had an appreciative audience, she added as soon as the sniggering had subsided: 'And the way she decks herself out, with flowers strung in her earrings, jewels around her neck, bangles on her wrists. Why, anyone might take her for a bride! Silly old fool!' said Gulab abruptly as if giving her final verdict, and now wanting the gathering to disperse.

'Yes, Gulab, we must get back to work,' said Champa, trying to bring quietness to the overexcited Gulab. 'It's no use rousing Unnie to greater wrath.' Anyhow, most of them felt they had had merriment enough for the time being and those who had already selected their bangles gradually sauntered into the house.

A group of Zohra's friends were constantly flitting in and out of her room, laughing and teasing her. The young matrons indulged in suggestive jesting which, without making her any the wiser, left her uneasy. Zohra did not like such grossness. Nevertheless, it added to the atmosphere, which like an intoxicant, stimulated her youthful blood with a vague anticipation, neither of pleasure nor of pain, but of something strange. Her heartbeats quickened and her colour rose and fell easily. Her companions misconstrued this as showing a happy state of mind. Even Rashedah, who knew Zohra's reluctance to the marriage, was deluded into the belief that she was now contented.

In the evenings, after prayer, everyone assembled in Zohra's room. The women-servants and maids sang the old songs from time immemorial which were sung at such festivities, Unnie usually playing the *tabla* and leading the choruses to ditties of which she had a great store. Thus the days passed quickly and happily, but at night Zohra often lay awake, musing for hours over her fate.

The spirit of revolt having passed, she had resigned herself to the idea of marriage. But her thoughts ran over and over again in the same vein: 'Allah alone can foresee what the future holds for me. My kismet will soon be sealed, for good or for evil, who can say? My parents have done their best to find a suitable husband for me and whatever manner of man he proves to be, I must make this marriage work. There is no retreat. Allah forbid that I should bring shame upon Abba Jan and Ammi Jan and upon myself by returning to their home disapproved by my husband's family. I would rather die than bring such shame upon my parents. It is they who have arranged this marriage, but I shall be held responsible for its success or failure. But, what if he does take a dislike to me?' This last possibility often startled and disturbed her. She knew that by Hyderabadi standards she was not beautiful.

With such thoughts for nightly companions, Zohra often rose in the morning, tired and listless; but the atmosphere of the house, with carefree voices ringing through its many courtyards and halls, always helped to dispel the gloom and revive her spirits. Zubaida Begum was anxious that the bride should look fresh. She had special sustaining sweetmeats prepared for her, even though Zohra had little appetite or inclination. But her mother's insistence was great and she had to force herself to make an effort to eat.

The idea of parting from Zohra was making the Nawab Sahib miserable. He often wandered into his daughter's room to see how she was faring. He would sit down and talk to her of Persian and Urdu poetry, quoting profusely. But marriage, or parting, was never mentioned between them.

At last the wedding day dawned and the house was astir with excitement. The maids rose and dressed early, in the new clothes the Begum Sahiba had specially had made for them for the occasion, and started preparing for the day's activities. They were anxious to make a good impression in front of Zohra Bibi's in-laws and other distinguished guests. Gulab and Champa wore pink tunics made from net, loosely falling over tight fitting dark green satin trousers and pink muslin dupattas, starched and crinkled. The new clothes gave them a feeling of festivity and a desire to show themselves at their best.

About eleven o'clock, there were wild cries in the zenana. 'The bridegroom has arrived. The *dulha* has arrived.' Everybody rushed to the window below which the bridegroom's procession was beginning to assemble.

Nalini, who was with Zohra, got up and placing a gentle and reassuring hand on her shoulder said: 'I must go and welcome my bridegroom-brother.' Zohra's heart started to beat violently; she felt a sudden impulse to go to one of the shuttered windows to see the stranger who would soon rule over her destiny. She rose but, even though she was alone, her courage failed her and she stood rooted to the spot whilst the awful fear ran through her: 'What if I should take a dislike to him? How shall I then have the strength to go through the ceremony? Alas, there's no retreat now.' Trembling and without having taken a single step, she sat down again.

Rashedah rushed in, exclaiming: 'Zohra, Zohra, your *dulha* has arrived!' Then, in a lower tone: 'There's no one about. Come and have a look at him. I'll bolt the door, so that no one catches you.' But Zohra only dropped her eyes in confusion.

'He's standing there,' went on Rashedah, looking surreptitiously through the window shutters, 'offering salutations to all. How tall and well built he is! Quite handsome. And there is Nalini, welcoming him. I envy her. But don't you want to see him?' Then teasingly: 'Of course, he's no concern of yours, but I am interested in my bridegroom-brother.' At this, Zohra's heart beat still more furiously. Turning round, Rashedah exclaimed impulsively: 'Zohra, how lovely you look! Shall I call Bashir Bhai to this window to take a look at his blushing bride?'

'Hush, Rashedah Apa, someone might hear you,' pleaded Zohra, almost in a whisper.

'There's nobody else but ourselves here,' said Rashedah and she launched a running commentary on the bridegroom and his people, certain that Zohra would be eager to know. Soon the bridegroom and his companions entered the house and, unable to see anything further, Rashedah came and sat by Zohra.

Zohra had been bathed with *chiksa*—a perfumed powder—and dressed in plain red clothes sent by her in-laws for the occasion. The bridal dress had to be fashioned in such a way that no scissors would be used on the fabric, scissors being considered a symbol of strife. The hand-torn dress was not shapely, but the all-concealing dupatta covered up the defects and fell in folds round her body. The simplicity of the dress accentuated Zohra's slim grace.

As the moment for asking the bride for her consent drew near, Zubaida Begum, Mehrunnissa, Rashedah, Nalini, and other women relatives gathered around Zohra and, from the men's quarters, came the Nawab Sahib and the bride's uncles. Her maternal uncle, her *mamoo*, was to act as proxy, and two paternal uncles as witnesses to her consent. They came up to her, and her *mamoo*, using the more formal form of address as befitted the seriousness of the occasion, asked: 'Zohra Begum, are you willing to be joined in marriage to Mohammed Bashir, son of Nawab Shaukat Jung Bahadur? If you are willing, will you accord me your proxy?'

No answer was forthcoming, nor was one expected immediately. It is customary for brides not to show unseemly eagerness by replying with immodest haste. Her uncle, aware of this, patiently asked again and again. The Begum Sahiba encouraged her daughter: '*Owi*, Zohra, how long will you make your *mamoo* wait? Say, *Bismillah*, in the name of Allah, and say Yes.'

Tears were flowing profusely over her face, dimming her vision. Her uncle's voice came to her as from an increasing distance, and her mother continued her coaxing. At last Zohra made an effort. She heard herself faintly answer 'yes', but it seemed like the voice of another, as her head felt dizzy. She had a strange feeling as if something were closing in upon her. The Nawab Sahib, deeply moved, placed his hand over her head and blessed her as did her uncles. The men all went out again.

Zubaida Begum wiped her damp eyes as she embraced her daughter. Mehrunnissa's eyes were also moist. Nalini was pensive.

The qazi was the man empowered under Islamic law to perform the marriage, and he waited patiently for the uncles to gain the bride's consent. At last they appeared, quite satisfied. A substantial sum was fixed as *mehr*—security money for the bride in case unforeseen circumstances led to a divorce. Under the colourfully decorated *shamiana*—an awning stretched on gaily draped poles and festooned with flowers which had been specially erected in the garden for this auspicious occasion—the qazi spoke the words. The bride was not present but the uncle to whom she had given her proxy accepted on her behalf. Zohra and Bashir were now married.

As soon as the ceremony ended the bridegroom rose and offered his salutations first to the qazi, then to his own and Zohra's father, then all around. He was warmly embraced by some, while others gave him their blessings and congratulations.

In the zenana, Zohra's relatives clustered round her and the felicitous sounds of '*Mubarak*', '*Mubarak*'—may it be auspicious—rang in her ears. Zohra made no acknowledgement for, on her wedding day, a bride is a beautiful and interesting model in the hands of others, with no action and speech of her own save with the most intimate friends and close family. In her present state of mind, this custom was a boon to Zohra.

The bridegroom's sister, Safia, accompanied by her mother and cousins, was ushered into Zohra's room. The bride was seated with the thick veil drawn well over her face. Masuma Begum, her mother-in-law, unsettled by the emotions that the occasion aroused in her, allowed her daughter to perform the ceremonies. The bride's people now stood aside and watched the proceedings. Safia, full of enthusiasm, lifted the veil and, raising Zohra's face on the palm of her hand, looked at her. Although Zohra's eyes were shut, she could feel the close scrutiny and vainly tried to lower her face again. Safia held it firmly, displaying it to the admiring gaze of her cousins. Then, sitting down beside Zohra, she took two strings of tiny black beads from the round silver tray held up to her by her cousin and fastened them round the bride's neck. This was a Hindu custom adopted by the Muslims and was believed to ward off the evil eye. The bride's people marked that Safia's movements were far from graceful. Safia then tried to push a large gold nose ring, with a ruby between two big pearls in the centre, into the hole in the left nostril, but finding it too small, she started pushing it into the hole in her ear. Masuma Begum interjected: '*Owi*, Safia, couldn't you have tried a little harder?'

'No, Amma Jan, or the nose would get sore,' said Safia decisively. The bride's people, who had watched with alarm Safia's rough ways, were somewhat relieved. But the bridegroom's mother complained: '*Owi*, what are these modern ways!' She sounded disapproving, for the black beads and nose ring were symbols of married status, the latter to be worn at least on the wedding day.

Then Safia put around the neck of her newly-wed sister-in-law an elaborate necklace of western design, which had lately come into fashion, a present from the in-laws. Safia had a passion for everything foreign and new, but she lacked the aesthetic sense to choose rightly. The bride's relatives, though they remained silent, looked askance at the ornaments.

A set of glass bridal bangles of exquisite workmanship overlaid with lacquer and gold and set with tiny pearls were handed to the bangle seller, who had designed the bangles especially for this occasion. She had come herself to put the bangles onto the bride's wrists, for this required a special skill. If they were to break they could not be replaced immediately. The bangles were small. They had to sit tightly on the wrists for to have soft malleable hands was a sign of beauty. As the bangle seller deftly manipulated the bangles over Zohra's hands they hurt her, slightly bruising her. But a wrist without glass bangles represented widowhood, and Zohra suppressed all cries of pain in the presence of her in-laws who were watching for any reaction from the bride. Her work done, the bangle seller turned to Safia and said in self-approval: 'Look, Begum, how lovely they are, like haloes round the moon!'

At last Safia and her companions left the room. But as soon as they had gone out, others came in a continuous stream. All were eager to see the bride. When the same people drifted in again and again, Rashedah became exasperated. At last, taking advantage of an opportune moment, she shut the door, exclaiming to Mehrunnissa: '*Owi*, have these people never seen a *dulhan* before? How pale and tired she looks already and there is a long ordeal yet before her.' Then, attempting to lift Zohra's spirits, she added: 'Let's see who dares to come in now. Not even the *dulha* himself will find entrance!' She knew well that no bridegroom would ever dream of entering the bride's chamber uninvited. Nevertheless she went on: 'But, Zohra will side with him now. All my years of devotion will avail me nothing. How unfaithful girls really are!'

'Pray, remember this when you're married, Rashedah,' said Mehrunnissa. Rashedah responded with a happy look of anticipation.

A weary smile came to Zohra's lips, as she realized the truth of Rashedah's words.

Mehrunnissa sat for a while, fanning herself and Zohra alternately, lazily holding the gold-and-silver handle of the embroidered fan. At last, rising, she exclaimed dramatically: 'Ah, how tiring! I am almost fainting with the heat. There is no time even to breathe.' Turning to Zohra: 'I must go and make sure that your in-laws have feasted well.'

'Yes, Apa Jan, make sure that they are in good humour,' Rashedah laughed.

'I'll send in something for you to eat,' said Mehrunnissa, 'Rashedah, please see that our *dulhan* eats well. She is in your charge today.' Mehrunnissa was feeling genuinely solicitous for Zohra.

'Allah, how my back aches!' Mehrunnissa, though fond of exaggerating her ailments, was really tired today. She was recovering from the weakening effects of a miscarriage. Zubaida Begum, anxious for her, had tried to persuade her to rest. But Mehrunnissa enjoyed the fun of the occasion and her own importance as the daughter of the house, and was unwilling to miss it.

Nalini brought in food on a tray for the bride. Zohra had little appetite, but Rashedah kept on pressing her to eat. Zohra was very quiet: there was a haunted look in her eyes. After the meal, Rashedah made Zohra lie down and, fanning her, tried to divert Zohra by carrying on a light, almost one-sided conversation full of innuendos and pleasantries. At last Zohra burst out impatiently: 'Rashedah Apa, can't you leave me alone even today?'

Rashedah looked at her in surprise. '*Owi*, what ails you, child? You look so frightened. One would think you were the only girl ever to get married.'

'Leave her alone,' intervened Nalini quietly. 'Marriage is after all a kind of dedication—dedication to another family—to a new person.'

'*Ai-hai*, this sounds like a sermon to me; "dedication" and all that!' But moving closer, Rashedah affectionately threw her arms round Zohra, who clutched her hand and burst into convulsive weeping. The tension once loosened, she wept until the dull relief of exhaustion came to her. Rashedah blinked away her own tears and tried to soothe her. Nalini stood quietly beside Zohra, but her deep eyes were disturbed. When Zohra had regained some of her composure, Nalini went out again, and Rashedah affectionately started to reproach her: 'Zohra, you are a fool!' Then more seriously, 'But what *is* the matter?'

'Rashedah Apa, what can I say? I feel a strange restlessness in my heart. Perhaps it is only my foolish imagination, but something oppresses me heavily. Maybe, I'm not made for marriage. I may not be able to make him happy. I myself don't know what it is. How simple life would be if one could swallow a diamond-chip and sleep for eternity!' Her voice was full of passionate sincerity.

Rashedah, unable to find an apt reply, could only say: 'Hush, don't talk nonsense, Zohra. No need to be frightened. Look at Mehrunnissa, is she not happy?'

'Rashedah Apa, how can I explain? I feel a deep foreboding, I know not of what. It cannot be death for, at this moment, death doesn't frighten me.'

'It is nothing but the stresses of the day,' said Rashedah, trying to be practical. 'Rest awhile, Zohra. Close your eyes and I'll fan you to sleep. You only need sleep and rest, and your foolish fears will vanish.' Rashedah, in her heart, was feeling perturbed at the sight of this sensitive overwrought girl, who sounded so desperately sincere.

In the zenana, all were feasted. First the mistresses and then their maids. After the meal, some of the ladies left, but a large number, including the bridegroom's womenfolk, remained to watch the *Julwa*—revelation ceremony—which would take place in the evening when the bride and bridegroom would get their very first glimpse of each other in a mirror.

The guests sat on the floor in the zenana on carpets covered with white sheets and reclined against long cushions placed against the walls and pillars. From the courtyard, a gentle breeze floated in, whilst green shrubs and the playing fountain made the blazing sunshine seem less intense. The ladies spent the long hours of waiting, in conversation, sometimes flashing into sparkling mirth. The pauses in the chatter were occupied in eating paan. There were several silver paan-boxes kept in different corners of the room. Besides, many of the Begums, addicted to the habit, had brought their own boxes with them. They prepared paan for others too and handed them round. The party was enlivened from time to time by the professional singers and dancers engaged for the day.

Outside, the men feasted under the colourful *shamiana*. There was a constant stream of arrivals and departures here, throughout the morning and afternoon, and food was served continuously to all newcomers. As each guest left, a souvenir of a small silver filigree box was offered with paan and a perfumed rose; and rose water from a silver filigree container was sprinkled on their clothes.

In the men's quarters, only the bridegroom's party remained, with a few of the more intimate friends of the Nawab Sahib. They were entertained by the singing and dancing girls, who were in their element here. They sang suggestive songs, and danced enticingly, casting sidelong glances at the more enthusiastic members of the audience, who showed their approval by applauding or, sometimes, when particularly appreciative, throwing silver coins. Outside the front gate, a row of beggars clamoured for alms. Food was generously distributed to all.

The long-drawn-out day passed, and evening approached. The time for the bride's departure was nearing, and Zohra's bridal make-up was redone, only to be washed away by her tears. But the gold powder sprinkled in her hair glistened still. The Nawab Sahib, the Begum Sahiba, Mehrunnissa, and other relatives gathered in the bride's room to bless her and bid her farewell. As each one embraced her, Zohra clung to them as if she could never tear herself away. Zohra wept as if her heart would break. She felt bewildered and deeply unhappy.

Meanwhile, in the centre of the zenana room was placed the bridal *takht* with its handsome silver legs. A cloth of crimson and gold was spread over it. This was to go with Zohra as one of the gifts from her own family.

Loud and excited cries of 'purdah', 'purdah', 'the *dulha* is coming', were heard on all sides. A long roll of cloth was brought out and held around as a screen for those who wished to observe purdah and yet wanted to see the bridegroom. But there were many who did not worry, for the bridegroom is a privileged person and no strict purdah is observed in his presence. Nevertheless, it is expected of him to have purdah in his eyes.

The bridegroom arrived, his sister Safia escorting him. Scarcely lifting his eyes, he offered salutations all around, which only the older ladies acknowledged with: 'May you live long'. With Safia for his guide, he removed his slippers and, mounting the *takht*, sat down cross-legged. Safia, as mistress of ceremonies, sat beside him; his mother and cousins moved closer but the bride's people, watching him keenly with their half-raised eyes, kept a bashful distance.

The bridegroom, tall and broad-shouldered, wore a long brocade *sherwani* with tight-fitting white trousers. On his head, he had a maroon turban tied in the fashion prevailing at the court of the Nizam. Strings of perfumed flowers, jasmine and roses, intertwined with gold and silver threads, hung from it in long streamers almost touching the ground. These streamers originally covering the head, were now thrown back over his shoulders, framing the face in flowers. Hushed exclamations of approval could be heard on every side.

The dancing girls took up their refrain to the accompaniment of music with great zest, and danced with gay abandon, their skirts whirling round their bodies. The noise they made was enough to drown all voices. The bridegroom, turning to his sister, said in a low tone:

'I look a fool sitting here, a cynosure for all eyes, and these dancing girls, I can't stand them; they are so unutterably vulgar!'

'Hush .... what does it matter? You can shut your ears. I can assure you, for once you look presentable,' replied Safia, taking a secret delight in her prim brother's discomfiture.

'This marriage business is quite a pageant. And I feel like the clown of the show.'

'On your wedding day, at least, you should conform to our ancient customs with better grace,' retorted Safia, feeling rather important at having for once the right to guide Bashir's behaviour.

'What with this turban and the heavy scent of these flowers, I'm feeling quite dizzy. Safia, why don't you hurry them up? It's stifling in here.'

Safia, lifting the fan lying beside her, said: 'I'll fan the bridegroom to keep him cool.'

'No, honestly, I feel like bolting.' Outwardly, Bashir looked composed, but Safia could see, from the way he was fidgeting with his hands, that he was nervous.

'Patience, Bhai Jan, patience. You know the saying "Patience begets sweet fruit", and you will see how true it is if only you wait a little longer,' exclaimed Safia archly. She had never felt at ease in Bashir's company but now she was experiencing a sort of liberation at being able to talk to her brother in this manner.

Soon cries of 'the *dulhan* is coming', 'the *dulhan* is coming' were heard. There was an eager turning of eyes and straining of necks to see her. She was supported to the *takht* well wrapped in her red dupatta and was seated opposite the bridegroom. Her head was bent low, the chin resting on a raised knee. A small silver mirror was placed between them, and a thick red covering was thrown over their heads, wholly enveloping them. After the recital of a verse from the Koran, they were to see each other for the first time through the reflection in the mirror. Although this time-honoured ceremony was always punctiliously observed, it was rarely possible to see anything clearly under such conditions. Besides, Zohra, startled and agitated, closed her eyes, and lifting her face from her knee, bent it still lower. Nor did Bashir make any real effort to look at her; but his hand, as if drawn by some magnetic power, quietly slipped towards Zohra's and touched the half-open palm lying in her lap. The soft hand was cold and trembling. With a sudden impulse he enveloped it in his own square palm, pressing it as if to give it warmth.

This first contact with his young bride made him suddenly aware that the ordeal was far more distressing for her than it was for himself. It would be in his power now to soothe or wound her, and with his chivalrous spirit newly aroused, he vowed with an impulsiveness alien to his nature that he would be kind and tender with her always. He squeezed more warmly the small hand, now so completely in his power; then quietly, though reluctantly, let go his hold, fearing that the covering would soon be lifted.

When Zohra first felt Bashir's hand touch hers, she was taken aback. Her natural impulse was to snatch it away, but instinctively realizing the futility of it, she allowed it to remain there listlessly. Besides, the grasp, though gentle, was firm and any effort on her part to unloose it would have ruffled the covering. This would have been noticed by the eagerly watching assembly and given rise to speculation and laughter, and made their position still more awkward. Already, quite unconsciously, she placed a barrier between the rest of the world and the two of them.

Zohra's frightened mind was somewhat calmed by this first contact with her husband. After the *Julwa*, the dancing girls noisily struck up their most suggestive songs, felicitating the bridal couple and their relations.

It was time for the departure, and Safia, watching her brother's reaction with mischievous delight, said: 'Bhai Jan, you must now rise and carry your bride to the palanquin.' Even Bashir gave a start, for he had fully hoped to escape this public ordeal. But the bride's people, proud of their heritage, insisted on observing all the time-honoured practices. Zubaida Begum, who had hitherto remained in the background, now moved forward to her daughter's side; Safia, making room for her, said to her brother: 'Here is your mother-in-law.' He rose and started to offer salutations.

But Safia exclaimed: '*Ai-hai*, Bhai Jan, bend down properly before her!' Whereupon Bashir, stepping down from the *takht*, moved closer and bowed so low that his head almost touched the Begum Sahiba's waist in the greeting used by the young to the elders and most respected members of the family. She placed her right hand over his head without actually touching it.

'May you live long!' Her lips trembled, as tears gathered in her eyes.

When Bashir straightened himself again, his mother-in-law passed her hands over him from head to shoulders in the customary manner, then cracked her knuckles against her own temples in token of taking on all possible mishaps from his head to her own.

Bashir was struck by the classic beauty and nobility of his mother-in-law's face. Although he had heard it much praised, he had not expected such perfection.

The bridegroom now looked for his slippers and found them missing. Guessing the cause, he helplessly turned to Safia, on whom he was so completely dependent today. Peals of laughter were heard from the bride's friends, and Safia, rolling her eyes and looking shocked, demanded: '*Owi*, girls, what is this? Where are my Bhai Jan's slippers?'

Rashedah, who had kept herself in the background, now came forward hesitantly.

'*Ai-hai*, Safia Apa, how can we allow our bridegroom-brother to carry his bride across the threshold until he has appeased us?'

'*Towba*, what sort of sisters-in-law has my poor brother acquired?' asked Safia with a feigned look of disapproval.

'Whatever sort they might be, it's too late to mend matters now. And unless our hands are warmed for sweets, he cannot take his bride home.' Rashedah stood close to Zohra as if to safeguard her.

Rashedah and Safia were in their element exchanging such banter, whilst Bashir, standing aside, was feeling more and more wretched. At last Safia, taking pity, said: '*Ai-hai*, Bhai Jan, there is no getting away from these sisters-in-law of yours. Produce your gold coins and be done with them.'

Bashir, who was prepared for some ransom, produced five gold sovereigns. Rashedah, accepting them, shyly glanced at him and said: 'Now, you have our leave to take your bride.'

The slippers were restored.

Most of the ladies now retired behind the curtain, as Zohra's father and uncles entered the zenana room. The bridegroom offered his parting obeisance to all. Helped by Unnie, he carried his bride to the palanquin, which had silver poles and a gold-and-crimson covering. The bridegroom himself then mounted a State elephant, lent for the occasion. It had a silver howdah lined with velvet-and-gold cushions.

The procession began to move slowly, headed by a band and lantern-bearers. All along the route, at short distances, rockets and various other fireworks were let off to herald the procession, and people ran out of their houses to watch it pass.

The stately elephant kept raising its trunk in salutation, whilst on the howdah, the bridegroom bent and touched his forehead in salaams. But he found the whole ordeal unpalatable.

The bridal palanquin was carried along on the shoulders of eight palanquin-bearers. They voiced happy, rhythmic exclamations, reciting the words of the poetess, Sarojini Naidu:

Lightly, O lightly we bear her along,
She sways like a flower in the wind of our song.

The procession also consisted of gift-bearers, carrying presents from the Nawab Sahib and the Begum Sahiba to their daughter. Women carried on their heads huge trays containing clothes and jewellery, with rich fabrics thrown over them. There was also silverware, including the bridal bed with silver legs, covered in crimson velvet, richly embroidered, and everything else her parents thought Zohra would need. Sword dancers and musicians joined the procession on the way, brandishing naked blades to the rhythm of the music to enliven the journey.

With all this pageantry outside, Zohra was left inside, in the suffocating closeness of the palanquin, with her new sister-in-law. To the young bride, her future seemed a mossy pool whose depths her eyes could not fathom. She sat quietly, with her head still bowed and the veil covering her face.

Safia, feeling affectionate and concerned, impulsively put her arm around her, and lifting her veil, drew Zohra's head on to her own shoulder. Zohra, touched by this in her present overwrought state, clutched at Safia convulsively, and burst into sobs. She had now no control over her weeping. Safia, distressed, tried to soothe her.

'Don't be so frightened. There are no strangers, only friends in your bridegroom's home. Our hearts have already gone out to welcome our bride.' Then with a still warmer embrace: 'Bhabi Jan, remember, I fell in love with my brother's bride first.' As Safia held her in a close embrace, Zohra felt her heart go out to this new sister-in-law. It sounded strange to be called Bhabi, brother's wife, but her changed status was already asserting itself in so many ways.

Touching her hand, Safia said: '*Ai-hai*, you are cold and tired. Bhabi Jan, for my sake, please rest. There will be endless ceremonies awaiting you when we arrive home.'

At long last the bridegroom's home was reached, and Bashir was able to dismount. The garden and house were ablaze with lights, in readiness to welcome the couple.

The palanquin was set down and the bride stepped out, helped by her bridegroom and Safia. The bridal couple lightly touched the head of a sheep, which was tethered to a tree, thus symbolically passing on to the animal any ill luck which may have accompanied them. The sheep was then slaughtered as a sacrifice, leaving the bride and groom purified and cleansed of all evil as they entered their married life together. The bride was carried to the bridal suite and was seated beside the bridegroom. There were more rituals yet to be gone through. The bridegroom had to carry out the rite of washing the bride's feet. This he did by pouring a little milk over them, from a silver ewer.

There was also the exchange of sweets between the couple. Bashir pressed the sweets into Zohra's tightly closed lips, whilst from the bride the sweets were carried to his mouth by Safia manipulating Zohra's hand.

This was followed by the observance of face-revealing. The bridegroom's parents were the first to come forward separately. They had taken no active part in any of the proceedings yet. Nawab Shaukat Jung Bahadur, a frail and ascetic-looking old man, gave his new daughter-in-law a talisman with his blessings. Safia tied this round Zohra's arm. Masuma Begum, an excessively thin erect woman of advanced middle age, fastened on Zohra's wrists a pair of ruby-studded bracelets, her usually stern face now displaying a faint smile as she gazed at the bride and took all evils upon her own head in the traditional manner by cracking her knuckles against her own temples. Other members of the bridegroom's family also came forward one by one, to see the unveiled bride for the first time and to make her the customary gifts of jewellery or silver or gold coins. Safia embraced her sister-in-law with great warmth as she slipped a ring on Zohra's finger. Yusuf, her husband, who was standing beside her, asked:

'What are you wooing her for now? Your mediations are no longer needed. Leave her to her bridegroom.' He laughingly cast a significant glance at Zohra.

'You can sit down and have a better glimpse of our bride,' said Safia simply, giving him a sprightly look. It was apparent that she was very much in love with her handsome husband.

Yusuf seated himself opposite Zohra with his legs carefully tucked under him at the knees, and, arranging the lapels of his sherwani, said:

'Dulhan, open your eyes and cast but one brief glance of indulgence at this poor supplicant.' He looked beseechingly, even though he knew the bride would never lift up her eyes. Zohra only lowered her head, closing her eyes still more tightly.

At last the family left the room, leaving Zohra with Safia, but only long enough to feast at the wedding dinner, after which they wandered back to view the bride, while Safia continued to lift up the bridal veil, revealing Zohra's face to admiring and eager relatives. Nobody seemed to have any sense of time.

Bashir, who was weary to the point of revolt, could contain himself no longer. Beckoning Safia aside, he said: 'Haven't we had enough of this medieval tomfoolery yet?'

'Bhai Jan, after all, a marriage is a marriage!' retorted Safia with emphasis.

'Is that any reason to half-kill the bride?' he asked, with marked disapproval. Then in a more conciliatory tone: 'For goodness sake, have pity on her! Why don't you let her retire? Do you want to break her back?' He was full of solicitude but Safia could detect the suppressed expectation in his voice.

'Her back is not a dry stick that it should break so easily.' She was feeling quite brave today in the presence of her awe-inspiring brother. With a mischievous smile she added: 'But why such sudden concern?'

Bashir, in spite of himself, could not hold back an embarrassed smile.

Then, suddenly becoming serious, she said impulsively: 'Bhai Jan, please always be good to her.' Her voice was anxious.

Bashir could only say: 'That's all right, Safia. I don't think you need worry.'

Thus Zohra, shy, troubled, and bewildered, entered her new home to begin a new life.

7

The following morning, after the bridegroom had left, Unnie eagerly entered Zohra's room. She had accompanied the bride, as an old and trusted servant. Her duty was to wait on Zohra who would be hesitant to order about the servants in the home of her husband's parents. She also had to keep Zohra's parents informed as to how the bride was faring in her new life. Unnie started to ask Zohra oblique questions, but the latter maintained a studied silence.

Most of the furniture in the bridal suite had been presented to Zohra by her parents. But knowing the bridegroom's preference for western ways, they had asked him to select it. He had chosen a simple, post-war western-style set, to which his mother and Safia had added rather incongruous accessories, making the room look a disharmonious mixture of East and West.

Unnie continued to make little significant remarks, whilst trying to help Zohra with the sari. Actually, she was only getting in the way. Zohra, though gently discouraging, had not the heart to ask her to desist. The sari was of sky-blue georgette, embroidered with silver sequins and edged with a silver fringe. Glancing into the long mirror, Zohra was startled by her own reflection. Never before had she worn such beautiful clothes, or looked so lovely. The weeping had not left any visible traces. The clothes had been selected by Bashir's mother from her trousseau and sent in with Unnie. There was also a tray of glittering jewellery lying before her, but she was reluctant to wear this and make herself more conspicuous. Unnie was pressing her to adorn herself, when Safia came in, with her hurried, uneven steps. Anxiously appraising the bride, she greeted her warmly. Zohra remained quiet, her eyes downcast. Finding the tray of jewels still untouched, Safia exclaimed: '*Ai-hai*, Bhabi Jan, a bride must look a bride. What, no necklace? No earrings? What is this?' Safia immediately started to deck her out. 'Now, Unnie, does not our bride look heavenly?' she asked, raising up Zohra's face and gazing at her exultantly, making it extremely difficult for Zohra to maintain her composure.

'*Owi*, Begum, I have been trying to persuade her to wear some jewellery but who will listen to an old woman nowadays? Yet she could not utter a word of protest before her bridegroom's sister!' Unnie complained, more pleased than offended.

Lowering her voice, Safia said to Zohra: 'Bhabi Jan, Mother would like to see her daughter-in-law adorned, as she feels brides should be.' To Zohra this seemed to be a hint that her mother-in-law liked to have her own way.

Safia had brought strings of jasmine flowers. She was weaving them into Zohra's long plait of hair, when Bashir entered. At the sound of footsteps, Zohra had instinctively raised her eyes, but meeting her husband's she looked down in confusion. She tremblingly lifted the edge of her sari thrown over the shoulder, and quickly covered her head. The silver fringe formed an exquisite frame round her dark head.

The shy look of alarm, the startled gesture, the graceful turn of the body, all appealed to Bashir's senses. He thought, whatever else might be said against the purdah system—and there was much—it certainly enhanced feminine appeal to a remarkable degree. His wife looked so alluring that a thrill of possessive happiness passed through him.

Bashir was moved by unaccustomed sentiments. Impatient to be left alone with his bride, he said something to his sister in a voice inaudible to Unnie. Zohra quivered, and feeling her face flush, she turned her head away.

Safia laughingly cast an astonished look at her brother, but his face betrayed little emotion. However, she found an excuse to take Unnie out, to the unconcealed disappointment of the latter. The bride and bridegroom breakfasted alone.

II

In the evening there was more feasting at the house of Nawab Shaukat Jung Bahadur. The guests consisted of members of the family and their more intimate friends.

This was the day when everyone could have fun and break with the formalities between the families of the bride and groom. A mock battle would take place and the bride and groom would start it.

After dinner, the bridegroom was invited to the zenana, where once again he was the only male member present. He was seated on the divan opposite his bride. Between them was placed a silver stand containing flowers, dried fruit, and sweets. These they would throw into each other's hands. Then they were to beat each other three times with ornamental flower sticks thus signalling the commencement of the family battle. The bridegroom's relatives were gathered round the divan whilst most of the bride's relatives remained on the other side of the sheet, held up to form a screen.

A volley of green leafy vegetables was suddenly aimed at the bridegroom from behind this screen. Instigated by Safia, he responded by lifting bunches from his tray and flinging them across. It was not a fair contest, as the girls from behind the curtain could see him and take aim, whilst he could only hurl his missiles at random. As the uneven battle proceeded, with light-hearted laughter from the bride's family every time the bridegroom was hit, Safia came to her brother's rescue. She went over to the screen and, lowering it, flung back the bunches being hurled from the other side, in playful revenge.

'Safia—you!' shouted Rashedah, 'You in the enemy's camp! You can't hit us!'

'Yes, I can. You cannot hit a man from behind the curtain. *Ai-hai*, if you want to fight, come out and fight in the open!'

'*Owi*, how can I come out?' replied Mehrunnissa half affectedly, half sincerely; for although thrilled at the idea, she was not accustomed to meeting strangers.

'If you don't come out, who will? The bride's sister can no longer take shelter behind the curtain!' exclaimed Safia.

'Apa Jan, I will go even if you don't; you are not going to observe purdah from Zohra's bridegroom, are you?' Rashedah reprimanded.

As they were still hesitating, Safia, wishing to bring them out, started to attack them fiercely exclaiming: 'What's this? You came out in his presence yesterday. Brother has not changed since then, has he? I shan't stop, unless you come out!' she exclaimed now with great relish.

As the volleys continued, Rashedah was the first to step out. Mehrunnissa followed her; then other cousins emerged who had had their parents' consent not to observe purdah from Zohra's bridegroom. The frolic was now in full swing, both sides flinging vegetables at each other with zest. Bashir, in spite of all his efforts, could not really enter into the light-hearted gaiety. The girls noticed this with secret resentment, but they also marked that whenever the flying missiles missed him, he immediately turned round to see that his bride was not hit. This somewhat softened their hearts towards him. At last they called a truce. Rashedah, going over to Zohra, whispered about the bridegroom's solicitude into her ear, causing Zohra merely to bow her head lower in confusion.

The previous evening Bashir had hardly noticed Mehrunnissa, but Safia now introduced her as Zohra's sister. He was a little disappointed in her looks, for he had heard praise of her beauty. Then he threw a furtive glance at the delicate figure beside him. Perhaps it was due to his westernized tastes, but to him there was no comparison whatsoever between the two sisters.

It is customary for the bridegroom to be particularly attentive to his bride's sister; he is permitted more liberties with her than with any other girl outside the circle of his own blood relations. Bashir, being temperamentally reserved, found this difficult, whilst Mehrunnissa, accustomed to being admired, resented his behaviour, looking upon it as a personal slight.

Zohra, as naturally becoming to a bride, had remained quiet throughout. Rashedah, now sitting beside her, felt her hand, which though still betraying nervousness, was not cold and frightened as on the previous night. Altogether, Zohra looked more composed. Rashedah was much relieved, for she had been haunted by Zohra's convulsive sobs and by the fear that there might be something real in her forebodings.

III

Zohra was gradually getting accustomed to her new life. One day, after Bashir had gone to work, she went, as was the custom, to the courtyard to sit for a while with her mother-in-law. Masuma Begum had a forbidding air. Her long, drawn face was wrinkled more than her years warranted. She had small keen eyes; her nose was long and sharp; her mouth was hard and firmly set; the protruding chin proclaimed her, in unmistakable terms, as one born to dominate. Her hair, a medley of white and black, was smoothly combed down on either side of the high but narrow forehead. As Zohra approached, she gave her a quick, appraising glance, then smiled as she acknowledged the low salutation, by cracking her knuckles against her own temples and blessing Zohra: 'May you live long!' The voice was strident, in keeping with her general personality but Zohra felt a warmth filter through the shrill tones. She sat down on the *musnud* at a respectful distance from her mother-in-law, who quickly folded a paan and offered it to her.

Zohra was not a paan addict; she allowed it to remain in her hand, intending to eat it later when she felt more inclined, but her mother-in-law marking this, said: '*Owi*, Dulhan, it is all well for unmarried girls, but it is not becoming for brides to have pale lips.'

Zohra uncomplainingly put the paan into her mouth.

As they sat there, Safia hurried in, and greeted her mother with a mixture of respect and a certain nonchalance, which Zohra observed did not actively please Masuma Begum. She kissed Zohra, more in the western fashion, and exclaimed: 'May no evil eye fall upon her, but, Mother, is not your daughter-in-law looking radiant today?' She looked in happy approval at her choice of a bride for her brother, causing Zohra to feel acutely uncomfortable. Then she sat down with apparent effort; her movements were angular, almost mechanical.

'Allah! How stiff my joints are! I rode ten miles yesterday, all the way to Golconda Fort. And all these weeks, what with these festivities, I had not ridden at all.'

'*Ai-hai*, these army officers!' exclaimed her mother brusquely. 'Now Yusuf has to teach his wife riding! There is no sense of propriety left in young people these days.'

Safia, ignoring the remark, kept up her chatter. But after a while, looking at her wristwatch, she burst out: 'How forgetful I am of time! I am still loitering here whilst I have to do some urgent shopping before lunch.'

'*Owi*, Daughter, you are always in a hurry,' said Masuma Begum, 'as if electricity had got into your feet! Why can you not rest? I am sure our bride also wishes you to stay.'

Zohra raised her eyes pleadingly. In the presence of her mother-in-law, diffidence still prevented her from speaking. Safia, who, for all her careless ways, had a warm heart, was touched by this appeal.

'I wish I could stay, but there is a party at the club next week in honour of the visiting vicereine, and I must buy a new sari.'

'What a craze you have for new saris, Safia! There are scores of saris that you can easily wear. How you girls waste money nowadays!' Masuma Begum's stern voice was a little softened in spite of the apparent reproof.

'*Ai-hai*, one cannot be wearing the same old clothes over and over again. I want something new, something striking, that will make everybody envious,' said Safia, who, though lacking in feminine appeal, had an irresistible fascination for adornment.

'What an ambition! Your bridegroom has given you too much freedom with his money. He indulges you completely.' The Begum wished to sound reproachful and she also wished to impress on Zohra that she disapproved of such extravagance. But whenever Masuma Begum reprimanded Safia, an unusual gentleness crept into her tone. She was her only daughter and the Begum was painfully aware that Safia did not have much of either beauty or brains. This lack of agreeable qualities was accentuated—even in her mother's eyes—in contrast to Zohra as she sat beside her. Safia's face had a frank attractiveness, but her figure and manners did not have much appeal. She herself, however, did not seem to be conscious of this. She prided herself on being a sportswoman and a real companion to her husband. Yusuf, in his good-natured way, humoured her in this belief.

'I must really be off now,' said Safia getting up hastily and arranging the folds of her sari around her with awkward gestures. As Zohra raised her eyes in farewell, she looked so forlorn that Safia said impulsively: 'Mother, let me take Bhabi Jan with me. Yes, Mother, listen to me, we shall be back by lunchtime.'

But the old lady retorted at once, sternly: 'You have no sense, Safia. How can a new bride go gallivanting with you? People are still arriving here to greet her. What will they say when they hear the bride has gone shopping?'

'Let them say what they will. It will at least give them a topic for conversation!' retorted Safia carelessly.

'You may do what you like. You belong to your bridegroom's house, but you are not going to spoil my daughter-in-law. Allah only knows what is coming over you girls. You have no sense of responsibility, no duty, no stability, no repose; no sense of anything but pleasure and pleasure-seeking!' Masuma Begum's voice was now bitingly condemnatory.

Safia, being in a hurry, merely said stiffly: 'As you please.' But whilst leaving she gave Zohra a sympathetic look and so violent a hand-squeeze that it was all the poor bride could do not to scream.

When Safia was gone, Masuma Begum started talking to Zohra. She was somehow moved to confide in her new daughter-in-law.

'*Owi*, Dulhan, Allah knows how worried I am on Safia's account. She has a handsome and carefree husband who, I hear, boasts that he took to the Army chiefly because he was born with the looks so perfectly suited to an army officer. "Others get commissions but I am God's chosen," he says. But he gives her the freedom to do what she likes, with the result that they are both entirely spoilt. There are no elders in the house. It is five years since her marriage, and there is no child yet, nor any sign of one coming. *Owi*, Dulhan, I have taken her to lady-doctors. I have got her talismans. But nothing has proved effective so far. Allah knows what a burden this is on my heart. All these years I have longed to hold a grandchild in my arms.' This remark she made looking significantly at Zohra and then went on: 'Allah knows, I would even give my right hand if only Safia were to be blessed with a child. I have grave fears; how long will her dashing bridegroom's affections last? And Safia worships him, even though they sometimes quarrel violently. I feel there must be something more permanently binding. You know of no one in your family who has undergone successful treatment, do you?'

Zohra, not accustomed to taking part in such conversations, could only shake her head while her lips framed a hardly audible 'no'.

'Your sister has two beautiful sons, hasn't she?'

'Yes,' she murmured.

'Allah be praised for that!' exclaimed Masuma Begum fervently.

Zohra did not like the turn the conversation was taking, for her mother-in-law's tone seemed to suggest she was thankful there was no barrenness in her family. Zohra did not yet realize what a great compliment Masuma Begum had paid her, by confiding in her about Safia and Yusuf, for Bashir's mother found it very difficult to give confidences. The conversation was cut short by the announcement of a carriage at the zenana door. Zohra heaved a sigh of relief, for in her family, these matters had never been discussed in front of the daughters and she was finding it disconcerting in the extreme.

Zohra had the midday meal with her mother-in-law. She noticed that none of the servants were asked to sit with them at meals, even when both of them were alone without Bashir. Here, a servant was treated as a servant, and not as a member of the family, as was done in her mother's house. Even Unnie, who had accompanied her to her new home, was never asked to sit with them—a slight the old woman strongly resented.

After the meal, Zohra retired. She looked forward to the freedom of her room, as a thirsty stranger to the oasis. Resting on the divan she started to read, but soon the book lay idly beside her as her mind wandered off into musing. The change from one phase of her life to another had been so abrupt, that to her it was like going to sleep in one world and waking up in another. She often thought it would not have been difficult if she had had only her husband to consider. In fact, at present he seemed so anxious to be considerate of her every wish, that she hardly needed to make any effort to please him. Only it was all so strange; but how long could she continue to accommodate her mother-in-law merely by going out and sitting in her presence, almost mute, and always overdressed? She had no say even in the selection of her own saris yet, and Masuma Begum often chose the wrong combination of colours, which hurt Zohra's aesthetic sense. But she wore the clothes regardless, and without a murmur.

'Allah be praised!' muttered Unnie, breaking in upon her thoughts. 'Chhoti Bibi, at long last you are allowed to rest!' Sitting down beside Zohra on the divan she pressed the bride's feet and tried to crack the toe-joints. '*Ai-hai*, how tired you must be, sitting there for hours with your mother-in-law!' she moaned.

Zohra knew Unnie wanted to draw her out, and to get her to complain against her husband's mother. She maintained a forbidding silence and pretended to resume her reading. In a different tone now Unnie continued: '*Ai-hai*, put that book away, Chhoti Bibi. What do you want with books even here? Lie down properly and I shall massage your back and feet. How they must be aching!'

'No, Unnie, I am not as tired as all that,' Zohra protested with a smile. 'Resting here is enough for me. Actually, *you* need rest more than I do.'

Unnie, cracking her knuckles against her temples in a fit of affection, said: 'Why, Bibi, may you live for hundreds and hundreds of years; you think of nothing but us poor folk; we, who are more worthless than your shoes. But still you think of us, just like your mother—may she be blessed with armfuls of grandchildren!'

Zohra felt annoyed with herself for having no control these days over the colour that so swiftly rose to her cheeks. Until now, it had always been 'marriage', 'marriage', 'marriage'; and now it was 'children'. Unnie continued: 'In your mother's home, unlike here, servants are treated as human beings.' Her old voice became resentful as she spoke of the difference between the two households, but Zohra cut her short by observing: 'People have different ways.'

'*Owi*, so you have already forgotten your mother and are siding with your mother-in-law?' There was reproach in Unnie's knowing eyes. Zohra made no reply. Then, as the debate would not be diverted into a more intimate and revealing conversation, Unnie said: 'Chhoti Bibi, while you rest I shall also stretch my back here. I don't like going to the servants' quarters now.'

'Yes, Unnie, sleep where you like,' Zohra answered affectionately, sorry she had been unable to satisfy Unnie's curiosity. But all the topics that Unnie wanted to discuss these days Zohra found vexing. Unnie stretched herself on the carpet in the verandah with her paan-bag under her head and was instantly snoring away.

As the time for afternoon tea approached, Zohra rose and started to get ready. Unnie, too, sat up yawning, ate a paan, and hurried to help her. Zohra, wrapping herself in an elaborately embroidered sari, said with a sigh: '*Ai-hai*, Unnie, I feel so foolish always having to dress up like this. When do you think I can wear something simpler?'

'*Owi*, Bibi, how indeed can a new bride do that? She must always look decorative.'

'Anyway, tomorrow is Friday and I shall be going to Ammi Jan's. How lovely that will be! It seems ages since I saw them. I shall then change into my old kurta, and feel free again.' The freedom she sought was not only from this rich apparel but also from the stifling atmosphere of formality of which the clothes were merely a symbol. Only she would never confide this to Unnie.

When Zohra was dressed she went back to her mother-in-law. Whilst they were seated together, Bashir returned from his work. His mother greeted him, but it was not expected of a new bride to make any gesture of welcome to her husband in the presence of the elders. Therefore Zohra remained silent, with her eyes cast down. Bashir, unmindful of old conventions, addressed her, but she could not utter a word in reply. Sitting down, he suddenly declared: 'Mother, I think it is time that Zohra went out. I want her to meet my friends and their wives.'

'*Owi*, get some sense!' replied Masuma Begum, astonished beyond measure. 'How can a new bride and bridegroom go out visiting together?'

'Why, do you think we need a chaperon? Unnie might be a good one. What do you say?' With an ironic smile, Bashir turned to Zohra. Zohra only pressed her lips together more firmly to suppress her amusement.

'This is the good that has come out of your English education,' retorted his mother. '*Ai-hai*, young people have no sense of modesty left. Whoever heard of a couple, married hardly a fortnight, going out together, visiting people!'

'Mother, I see no reason for you to be so upset. We carry this modesty business too far. I am sure Zohra is tired of all this idleness and this formal behaviour. Why cannot she go out with me alone? After all, we are married,' argued Bashir in his matter-of-fact way.

'Do what you like,' said the mother bitterly, knowing how stubborn her son could be when his mind was made up. 'You can go and make an exhibition of yourselves!'

'You need not worry on that account. We shall behave ourselves,' he said, affecting not to understand her. 'Perhaps we shall go to Safia's. I know you will approve of that.'

'You can do what you please. Elders nowadays exist only so that their wishes can be flouted by their children.' Masuma Begum's embittered nature did not allow her to concede anything graciously.

Bashir, aware of this, did not argue with her any further. But he did not intend to give way either. He turned to Zohra, who sat there, listening quietly: 'Let us go. I know Mother does not really mind.'

Zohra did not know what to do. Confused, she turned to her mother-in-law in mute questioning, her eyes barely raised. This gesture somewhat softened the old lady's heart as she realized that at least her daughter-in-law was anxious for her approval.

'Yes, Bride, go. When your bridegroom wants you to go, who am I to prevent you?' The taunt was directed more against her son than his bride.

As the couple entered their suite, Bashir put his arm round Zohra, drawing her to him.

'I thought we were going out,' Zohra spoke, trying to draw back.

'Yes, but where is the hurry?'

Zohra, still keeping her distance with an outstretched arm, asked in a low voice: 'Is it right to displease Mother in this way?'

'Surely we cannot cling to old ways that have no meaning,' he said, leading her to the divan.

'But I am too new in this household to try to change your mother's beliefs,' she countered as they sat down.

'The slowness of things here!' He sounded impatient. 'I do not have a thousand years to live, and a hundred times to marry. Life is much too short to waste on foolish formalities. What we need is a general upheaval.' Then releasing her hand and speaking seriously: 'We need a revolution of thought, revolution of methods, revolution of systems. In Europe, science advances so rapidly that people are used to quick changes, but here we resist new influences.'

'But isn't Mahatma Gandhi doing that?' she asked, quietly moving a little away from him. 'Only, his is a new way of revolution. But it will surely lead to the regeneration of India,' Zohra added hesitantly, her fingers playing with the fringe of her sari.

'I, too, was of that opinion when I was about your age. I was in England then and longed to be back to do my bit. But frankly, I have never had faith in Gandhiji's methods. Metaphorically speaking, and in some cases literally too, we live in an age of aeroplanes and high-speed cars, but continue to travel cautiously by bullock cart in the absurdly leisurely manner reminiscent of prehistoric days. This is the Mother India of whom our poets sing in high-flown terms.'

Zohra was still too shy of her husband to argue. Besides, she continued to hold him in awe. Switching back to the original subject, she said: 'But how can I stay with Mother if I start displeasing her? Please don't make things awkward for me.' Her low voice was pleading.

'Would you not rather go out with me than be sitting here, submitting to Mother's visitors like a show-window model?' he asked, unexpectedly hurt.

'You well know that there can be nothing more embarrassing than being on display,' she answered simply, taking a deep breath.

'If Mother continues to want to dominate our lives, we shall go and live somewhere else. There has been enough unhappiness in this home.' Bashir's voice was now as hard as his mother's.

Zohra wondered whether the expression of sadness that often flitted across Masuma Begum's face had anything to do with it, but it was not her place to ask and she merely said: 'Is that not a reason for you to wish to bring some happiness into their lives? It would be selfish to leave Mother alone. Also your father, detached though he is from life, likes to feel the presence of his children about the house.'

'The obligation, at least, is on both sides then,' he said drily.

'We must find some way by which we do not hurt Mother's feelings. We must avoid dividing the house. I think one does owe one's parents a great deal.' She was reminded of her own parents now.

'This idea of eternal obligation! I am sure I shall have pleasure enough in my children without expecting such everlasting gratitude from them.' He tried to meet her eyes, but she looked away instinctively.

Bashir, putting his arm round her, drew her close to him. But, seeing she was pulling away, he lifted up her face and, gazing into it, asked in as tender a voice as he could command: 'I suppose we shall want children some day—not many—but a couple, at least?'

Zohra refused to meet his eyes and inclined her head to avoid them. Bashir knew it was not yet the time to pursue this subject. He knew also how difficult it was suddenly to indulge in any sort of passionate love-making. But she looked so utterly desirable; her lips, her eyes, her hair. With an effort, he restrained himself and, taking her hand in his own, asked: 'Look, but even if you weren't married would you not have wanted some freedom from parental control?'

'Which girl would not?' For the first time, there was a hint of defiance in her voice. Zohra's mind flashed back to the time when she had so intensely wanted the freedom to delay her marriage without causing grievous hurt to her parents. It all seemed so far away now, and was she not happy? She admitted hesitatingly that she was. Bashir was taken aback by her vehemence. Releasing her hand and moving away, he gave her a quick piercing look. Suddenly, he felt the need to know more about the heart and mind of his youthful wife.

'I think, Zohra, in a way you are right. Mother has suffered greatly,' he said in his deliberate voice. 'But how on earth can I get to know my wife under the maternal eye? We must get away from here, even though it be for a short time.'

'But where, and how?' she asked.

'It is only a month now to the summer vacation. It would be lovely to go to some hill station. Mountain air is so bracing.' He already sounded decided.

'That sounds delightful, I have been to Ootacamund. It was heavenly.' Zohra was trying to suppress the excitement in her voice.

'Let us go to the Himalayas. Mussoorie, Darjeeling, Naini Tal, or would you rather go to Kashmir? What do you say?'

'It is for you to decide,' she said quietly, not sure of herself yet.

'Any of these places is good enough for me. We can also explore around a bit.'

She could not resist the temptation to say: 'Then you had best join the Everest Expedition!'

'I might have done it, had I not married you.' His tone seemed to suggest that, if he had married somebody else, he might still have done so. He continued: 'I think, Zohra, the European system of going away on a honeymoon is most sensible. Here, we merely get involved with relatives and more ceremonies, formalities and false modesties.'

'But you forget,' she faltered, but then continued, 'their whole system is so different. After all, they know each other beforehand.' She still avoided using the word 'marriage'. She recollected to herself how nervous and frightened she had been, not knowing what the future might hold for her in a strange house with a strange bridegroom.

'Yes. Our system is hopeless. We move in reverse gear. Sometimes it may prove disastrous. I am surprised that it is not so more often. Everyone surely does not have my luck.' It required an effort on his part to utter this last sentence, but he wished to convey his feelings to Zohra without having to make lengthy professions.

Zohra's smile was bashful as she said half-teasingly, half-earnestly, 'You have not seen other girls.'

'I do not think it would have made any difference.' It surprised Bashir that he was able to express these sentiments to his wife, having married the way they had. He knew she could not realize how greatly she attracted him in every way.

8

The idea of the bridal couple going away to the hills was not welcomed by the elders. How would the bride manage, how could they go to an unknown house, were the sort of questions raised. But Bashir, once determined, overcame all opposition. He also refused to take Unnie or any of their own servants. He knew it would be impossible for Zohra to be natural with him, under their prying eyes.

Mussoorie was selected, because a friend of his offered him a furnished cottage there, with arrangements for servants. This simplified matters. Before they left, Zohra wished to spend a few days with her parents. Bashir, watching her preparations, observed: 'You look so happy. I hope you have a nice time.'

'Oh yes, I know I shall,' came the joyful reply, but marking his lips close firmly, she quickly added: 'You too will have more time to get through your work if I am not here. I shall be back soon. Ten days is a short time.'

'Anyway, I shall be seeing you at your mother's. I can drop in on my way back home from college,' he tried to sound casual, not wanting to betray his disappointment at her departure.

But Zohra gave a start. 'No, no, please don't! It is not becoming.'

'What is not becoming?' he asked. 'Or is it that you want to escape from me altogether and dislike even the idea of seeing me at your parents' home?'

Zohra shook her head, 'It is not that. It's just that it is not customary for husbands to run after their wives when they are visiting their parents.' There was a bashful note in her voice. She had never yet referred to him as her husband.

'Then it is a way of escape,' he stressed.

'If you like to put it that way,' she said in a low tone that held a hint of guilt, 'but I think it is a good custom. After all, I should not like to think that there has been a break with my old home. For a brief period, at least, I should love to be back there as a daughter who has no interests outside it. Your visits would break that spell of quietness.'

'In short, you do not wish to see me at your parents' home. I understand.' He tried to sound aloof to hide his dejection.

'Is it fair to take it that way? My parents have respected your privacy. They have never once come to intrude upon it. Now you should have respect for their home-life, and not break in upon it when their daughter is visiting them.'

'But I wish your parents did come sometimes. I should like them to come,' said Bashir.

'You know among our people a girl's parents would never burden their son-in-law in any way. They will invite us a thousand times but never come themselves. They will give us the most expensive presents, but never accept anything from us.' Then thoughtfully, 'I don't know, but perhaps it is a good convention. I certainly should not like you to feel that my parents were proving themselves a nuisance to you.'

'Which means that you or they want to treat me always as a stranger.'

There was a momentary pause. He broke it by saying: 'So it is final that you go into purdah again, from me at least?'

'Yes, please.' She smiled at him.

But, as he still looked puzzled and unhappy, she continued timidly: 'Besides, you know my friends would laugh and tease me if you came every day. You do not know how embarrassing that would be!' She drew in a deep breath, as she recollected the last few months. She wondered how she had survived them. 'But I know Abba Jan will invite you to come as often as you like, and you can come sometimes.'

'Thanks for the permission.' He could not keep the sarcasm out of his voice. Then, changing his tone: 'I had even thought I might come and take you out for a drive in the evening sometimes.'

'Please do not be hurt,' she said softly. Moving closer, she laid a diffident hand on his shoulder. She had never made any kind of approach on her own before. Catching hold of her small hands, he drew her fiercely to him.

II

A fortnight in Mussoorie had helped Zohra overcome her shyness of her husband to a great extent. In the grandeur of their surroundings, all seemed natural. Thus Zohra, her fears dispelled and no longer hampered by conventions and family obligations, settled down to married life, and was fulfilled. The mountain air had heightened her golden complexion to an exquisitely warm tint.

Their small cottage, with its red-tiled roof, lay in an arbour of green trees away from the main hill. A creeper of wild red roses caressed the walls on one side.

The living room had a big bay window overlooking the green slopes, slanting down into the valley. Zohra would sit there, watching the clouds rise and float away among the hills and in the valley, ever taking on new shapes.

Frequently she lay awake, long after her husband had gone to sleep, filled with the sheer ecstasy of the beauty around her.

Sometimes, couplets or quatrains formed themselves in her mind, and she felt the need to rise and write them down, for often they vanished with the night. But she had to be careful not to disturb Bashir. She remembered, too, how scornful he was of poets and poetry, and feared his ridicule.

One morning, as it was drizzling, she reclined on her favourite window seat, writing to her parents, whilst Bashir sat reading in an armchair. Zohra, having finished her letter, cast a glance at her husband. Seeing him wholly absorbed, she watched him carefully. His keen eyes were fixed on his book to the exclusion of everything else. Taking up the writing pad again, she started to sketch, observing him properly for the first time, his high forehead, from which the hair already showed signs of receding, his firm protruding chin, the deep-set eyes, keen and intelligent, the bushy eyebrows, the sharp nose and somewhat thick, passionate lips. His features were sharply pronounced. When Zohra had nearly completed the sketch, he suddenly closed the book and turned to her:

'Have you finished your letter?'

'Yes.'

'Then, let us go for a walk. The rain has stopped.'

'Just a moment.' And as she went on with her drawing, with a stealthy glance at him, he got up and went over.

'What are you doing?'

She placed her hand across the paper.

'Is it a report on me to your parents?' he asked, momentarily shedding his seriousness.

'Why not?' Her large eyes twinkled.

'It is fair that I should know my faults,' he said laughing. Then lightly leaning on her shoulder, he lifted Zohra's hand off the sketch.

'Why, I did not know I had an artist for a wife.' There was pride in his voice, and the pressure on her shoulder gently warmed her. 'But where did you learn?'

'I used to take lessons at school. I am not an artist though.' She did not feel worthy of that word.

'You have a gift. You might have made a living by it.' Zohra gave a start. Among the families of the Nawabs, professional careers for girls were not contemplated. She was disappointed that this should have been his first reaction. Bashir, unaware of this, went on: 'Why did you not go on with it?'

'Because marriage was thought to be more profitable,' she retorted with malicious humour.

Reminded of an unhappy chapter, her mind wandered back to her old dreams. She started to add to the sketch. Bashir, watching her, asked: 'Now, what are you trying to make of me?'

'Wait and see.'

Standing beside her, he watched her pencil move busily. 'You are giving me the Kaiser moustache! Do you want me to grow one?'

Holding it away from her, so that they could both see, she said: 'No,' and quickly rubbed it out with the other end of her pencil. 'But let us try a different kind.' She put in some dark shading in the centre, just below the nose.

'Is this not like Charlie Chaplin?' she asked.

'You want me to look clownish!' He seemed amazed for he liked to appear dignified.

'Can the mere addition of a moustache make a clown of you?' she asked, glancing up at him. 'But wait, I shall now make you look most dignified, with a beard!' Shading in a closely-trimmed beard, she asked: 'How do you like this? Or do you prefer the long kind?'

Absorbed in her hobby, Zohra laughed happily. Now that Bashir knew of her love for sketching, she could easily indulge in it, she thought.

Bashir felt more light-hearted and happy during these days in Mussoorie than he ever remembered feeling before. Marriage had its distinct rewards and they were happy in themselves, in their walks in the hills, as well as in their different pursuits. Time passed easily without much desire for outside company, at least on Bashir's part. Zohra was sometimes wistful, thinking of her friends. Bashir, twelve years her senior, at times appeared a little too serious.

Whenever possible, Zohra liked being at home at sunset for the evening prayer. Bashir, at first, tried to laugh away her scruples. But noticing she was genuinely upset, stopped making irreverent remarks, and got out of her way whenever she wished to pray. Zohra invariably spread her prayer rug in front of the window, in order to get a view of the sky.

One evening, after prayers, she went out into the garden and sat down on the bench, contemplating the iridescent clouds, floating in wispy shapes, against the burnished orange and sapphire blue of the sky. She gazed at it spellbound.

Bashir wandered outside and called: 'Zohra!' Then seeing her, he came and sat down beside her.

'Are the colours not spectacular?' Zohra asked in wonderment.

'Yes,' replied Bashir. 'It is a gorgeous sunset, but, of course, dust and impurities are responsible for that.' This scientific observation was stated in a matter-of-fact tone. For Zohra, the enchantment was instantly dispelled. She made no reply. Looking at the mountains he continued: 'There are vast stores of raw materials in these hills, but little has been done to mine them. We are hopelessly backward in these matters.' His disdain was evident. 'We are so obsessed with our souls and we envelope ourselves in mysticism.' He gave a slight shrug.

'Surely the soul is more important than the things you speak of?' Zohra's susceptibilities were hurt.

'India has done enough delving into all that, and where has it got her? I have no patience with so-called mystics. Let them do something worthwhile.'

'Do you mean that these things are not worthwhile?' she asked a little defiantly.

'Look at the hordes of fakirs and sadhus that descend upon us, begging in the name of asceticism. Can you tolerate that?' he questioned back, piercing her with an intense look.

Zohra answered: 'The real yogis, you well know, live far away in isolation in ashrams or in solitary caves, seeking affinity with the divine.'

'It is only a way of escape from reality, a form of cowardice.'

'To me it shows a desire for a nobler life.'

'Mysticism and spirituality are actually states of superstitious ignorance. You must keep your mind clear, Zohra, and not get confused with such ideas.' Zohra recognized the condescension in his voice and resented it. He continued: 'Science is real; it gives one the right perspective.'

'Does your science teach the way to a higher life—the way to save one's soul?' she asked, struggling to keep calm.

'To a higher life? Yes, certainly. But what is this soul, for the salvation of which we sell our comfort and our happiness? If it existed, surely we should know it without having to renounce the world and going in search of it. I must confess I have never felt any such urge.'

'If you have not, you never will. Look at Gandhiji and his soul-force!' She gesticulated forcefully, as she became excited.

'Yes, we are bringing this soul-force into our politics too!' he commented drily.

'Oh, it is no use arguing,' she said heatedly, her eyes flashing with mingled sorrow and anger. Hurriedly rising, she walked into the house.

Bashir followed her after some time. He quietly seated himself on the arm of her chair, where she sat brooding. He started gently to caress her. He felt easier with such expressions of his love, rather than with verbal endearments. Bending down, he inhaled the fragrance of her hair. As she remained motionless, he whispered: 'Forgive me for causing you pain. I did not mean to.'

'Yes, I know.' She was subdued. Quietly freeing herself from his attentions, she said: 'But you frighten me sometimes. I feel, oh, I do not know how to express it, but I feel that, were it possible, you would place beauty itself on the dissecting table and not feel it a desecration! Oh it is difficult to explain, but you do understand what I am trying to say.'

'Let us forget that now!' He brushed her hair with his lips, then looking out of the window: 'The moon is beautiful. Let us have supper, and go out.' He wanted to restore her to a happier mood.

III

Zohra and Bashir had been in Mussoorie for nearly a month, when one day Bashir accidentally met an old friend, whom he was glad to see.

Shafqat was a deputy collector in the Punjab, and had come to Mussoorie on a month's holiday. His wife, Jehan Ara, was of the voluptuous type. She had a creamy complexion, which was usually hidden under make-up. Her features, though heavy, were well proportioned. By all standards of feminine beauty, she was a handsome woman. Shafqat, although an intellectual, was already falling into the smug inertia that so often overtakes the civil servant in India.

Zohra gathered from Jehan Ara's casual remarks that she had, of her own free will, married Shafqat for the glamour of the civil service. This shocked Zohra, for she had romanticized marriages of free choice, believing that, if girls had the freedom, they would marry only for love.

The couple had come with the professed intention of having a good time in Mussoorie. Therefore, when Bashir and Zohra met them, it meant a change from their own quiet mode of life.

Under her loud and flirtatious ways, Jehan Ara had a warm heart. Detecting a kind of loneliness in Zohra, she voluntarily started counselling her on all aspects of married life. Zohra listened to her with a shy deference.

During one of their very first meetings, leaving Shafqat and Bashir absorbed in old reminiscences, Jehan Ara had walked away with Zohra and, sitting down on the edge of a low stone wall, had started to talk to Zohra in a confidential manner. She seemed to think it her mission to advise all younger and less experienced women. When Zohra asked her about her child, she said: 'You know, I have one child, and he is mostly with Mother. Small children are such a nuisance.'

'But you must be missing him,' observed Zohra.

'Oh no! He's happy with Mother, and she is happy with him. So, why should I worry? Children are such a responsibility. Catch me having another!' she ended flippantly.

'But, won't he be rather lonely?' asked Zohra, amazed at an attitude so novel to her.

'He'll soon be going to school where he'll be all right. Why, you sound as if you intend to have a dozen and set up a whole colony at home!' Jehan Ara's rolling eyes inquisitively scanned her all over.

'No, no!' exclaimed Zohra hurriedly, aware of her own innocence in these matters.

'One has to consider one's figure,' Jehan Ara added, passing her hands approvingly over her sides as if to show off the full curves, with which she could afford to take no risks.

Zohra's surprise increased. This was her first insight into a fashionable woman's life.

'We, that is, my family, discarded the purdah long ago and, thank God, we're free!' declared Jehan Ara, proudly pulling herself up to an erect posture. Then she continued: 'My brother is soon coming to join us. I'm sure you will enjoy meeting him. He's full of fun, and just the right companion for a girl of your age. He is twenty-three, two years my junior.'

Her brother, Siraj, arrived. He was handsome in a flashy manner, and no peacock could have spread out his feathers before his favoured hen more than did Siraj before Zohra.

Once, as Bashir and Zohra were taking one of their favourite walks, they encountered Jehan Ara, Shafqat, and Siraj. Bashir felt resentful as there was no alternative but to join them. But he soon became absorbed in a discussion with Shafqat, whereupon Jehan Ara burst out: 'Oh, their eternal polemics and arguments! From the seriousness of their demeanour, one would think they had to settle the fate of the whole world. Zohra, let's go ahead.'

Zohra cast a glance at her husband. Seeing him in deep conversation, she readily agreed.

Suddenly she felt light-hearted and carefree. When they saw a stream flowing down the hill, Siraj exclaimed enthusiastically: 'Let's follow this stream to its source. They say it flows from a rock shaped like the spring-demon's mouth.'

'It must be the elixir of life then. Go ahead, Siraj, and drink of it for eternal youth; and you too, Zohra!' said Jehan Ara, tossing her head, even though there was no one around on whom she could exercise her charms.

'But what about you, Apa?' Zohra could not bring herself to call Jehan Ara, who was older than her, by her name.

'Oh, I am too tired to move another step. I shall rest here a moment and then go back to the house. Tell Shafqat, when he arrives, that I've gone.' Zohra was getting accustomed to Jehan Ara calling her husband by his name in the western style. In Hyderabad, it was a thing so rare that she had never heard it yet.

'Let's move on,' said Siraj.

'Should we not wait for the others?' asked Zohra. She was not used to walking alone with young men, and she was also worried that her husband may not approve.

'What for?' asked Siraj.

'We are so far ahead, they might miss us.'

'Oh, they will follow us all right. Bhai Bashir has the sure instinct of a beast in the jungle who unerringly follows his mate.' Zohra did not like this remark at all, but Siraj's smile was so disarming that she could not take offence. Bashir, she knew, was a studious and accomplished man, not given to frivolity. In the presence of Siraj, Zohra felt like a young girl again.

They walked along the edge of the stream and like children, jumped over rocks in their path instead of skirting them. She wondered if Bashir would consider them silly.

'Be careful of your sari! Here, I'll help you over the rocks.' As he took charge, Zohra did not know how to object.

'I am sure I can manage for myself,' she protested, determined to repel his increasing attentions.

Siraj plucked golden daisies and handed them to Zohra. She accepted them and, not knowing what to do, quickly tucked them into her hair.

'Oh, no, not like that,' he laughed, 'they should really be a setting for your face. Here, I'll do it.' He moved closer and pulled at the flowers.

'No, no. I can do it myself.' Feeling uneasy, she snatched them from him and rearranged them in her plait.

Staring at her in open admiration, he burst out: 'They look lovely, although I could have done it better.' He tried not to show his disappointment at not being able to touch her hair. 'But anyhow, if you'll permit me to say, you look enchanting. Golden flowers in raven tresses! The poet's imagination has always been carried away by flowers entangled in the snares of a woman's hair. Only, more things than flowers get thus entangled!' He gave a meaningful look, as his hand passed over his own heavily pomaded long hair.

Zohra, glancing back in confusion, could only say: 'Those two are long in coming.'

'They are engrossed in wise deliberations. Let them take their own time. We needn't wait for them.'

The depth and width of the stream often varied. Against her will, Zohra was finding this walk with Siraj exhilarating. Looking out for a diversion, she noticed two rocks close to each other.

'I shall cross over to the other bank of the stream,' she said, 'and race you.'

'Why, are you afraid of me?' he asked, raising his eyebrows and giving an engaging smile.

'Oh, no!' protested Zohra, ashamed of her own insincerity. But, before he could stop her, she had stepped onto one of the stones and crossed over to the other side.

They ran along either bank of the stream, laughing at each other, she nervously, he exultantly. After a steep stretch, she stopped for breath, and Siraj, spying two more stones, said: 'Let's change sides again.'

Without waiting for a reply, he stretched out his hand and, almost involuntarily, she put out hers to meet his. Each stepped onto a stone, and with a swing of their bodies, crossed over simultaneously. After this, whenever Siraj sighted two stones close to each other, he stretched out his hand, and with each contact Zohra found the pressure of his hand becoming warmer, more intimate.

At last they reached the source of the spring. They were thirsty and, cupping their hands, they started to drink the fresh bubbling water.

'Your hands are so small. Drink from mine.' Siraj held out his hands to her mouth. Zohra drew back, now really alarmed, her eyes flashing.

'Oh, Zohra, what a child you are!' he exclaimed, laughing at her; then he began to sing:

> *Fill the pitcher, fill the pitcher,*
> *Fill it, till it spills.*
> *Dawn is glinting, earth imprinting,*
> *Tinting every hill*
> *And pinking every rill;*
> *Bangles strike on earthen pitchers*
> *Music that can kill.*

Our husbands tell us
They are jealous
For they know it will
Make other bosoms thrill.

The tune was catching.

Siraj was so irresponsibly charming and was such a change from her serious husband that she could not help being attracted. But as the excitement mounted, she felt the situation was likely to get out of hand, and she would not know how to handle Siraj. She wished Bashir would arrive soon.

'I like this melody, do you?' Siraj asked, having come to the end of his song.

Not knowing what to say, she half raised her eyes to his. He tried to hold them but she turned away.

Feeling both physically and mentally exhausted, she sat down on a rock. Siraj came and sat beside her. Touching her bangles, he asked: 'Is this Hyderabadi workmanship?'

'Yes.'

'It's exquisite.'

Gradually he slipped his hand on to hers, covering it. 'I'll take care of you. You are so frail, you may slip down into the water.'

Zohra tried to draw it away, but he held it firmly, saying: 'Why, let's be friendly.'

'We can be friendly without this.'

Siraj only laughed at her discomfiture: 'Is Bhai Bashir as possessive as all that?' His look expressed a great deal.

'Please let go of my hand!' She tried to jerk it free. Siraj released it.

Still unabashed, he asked: 'Tell me, what does it feel like to be married?'

Zohra avoided his eyes. Siraj, watching her, was spurred on:

'Exciting? How nice to have someone so much in love with you. I've never had that experience. I think I am not fated to. My kismet!' He forced a sigh.

The self-assurance in his voice belied his words. He was conscious of his own power, being no novice in the art of flirtation. As Zohra did not respond, he again burst into one of his light-hearted songs:

> *We drink and make merry,*
> *With drinking we bury*
> *All thoughts of depression and sorrow;*
> *For life with its trouble*
> *Is only a bubble*
> *That's sure to blow over tomorrow.*

He sighed again.

Zohra, to break the spell of her own silence, said: 'These are Harin Chattopadhya's songs, aren't they?'

'Yes,' he replied glibly, not having the least idea of their composer's identity.

'He used to come to Hyderabad to his sister Sarojini Naidu's house. What a gifted family! They say he is a born actor.'

'I love acting,' he broke in unceremoniously, carried away by his own enthusiasm. 'How I wish Father would let me join the films! We have such poor cinema stars.' He seemed so sure of his own histrionic talents that it made Zohra smile. But that very instant it struck her: was he not already trying to look like a star? She had been feeling that his mode of dress was a little too showy. But Siraj in his zeal, with a sweeping gesture, flashing his sapphire and ruby ring, continued, 'Look at the standard of American films. They're superb. Do you like them?'

'No—that is, very few. But I have not seen very many, and I suppose I am not used to them yet,' she said, not wishing to contend his judgement.

'There are many things you'll enjoy when you get used to them,' he said with a significant look, making every attempt to provoke her into retaliating.

Zohra swiftly jumped to her feet, saying: 'Let us gather ferns.'

'Oh well, if you wish it,' Siraj said peevishly. Having no alternative, he got up.

At last the two men arrived.

'Hello, you two, and where is Jehan Ara?' asked Shafqat.

'Oh, she was tired and decided not to wait for you,' Siraj replied.

'Why did you disappear so fast, Zohra?' To Zohra's sensitive ear Bashir's even voice seemed to carry a note of anxiety. 'We almost lost trace of you. It was only my instinct that brought us here.'

Siraj cast a glance at Zohra, and though her eyes were turned to her husband, she was aware of this look, but avoided meeting it.

IV

Bashir disliked Siraj instantly and was distrustful of him, but he argued with himself that it was absurd to be worried about the first young man Zohra had met. But then, he knew she had led a secluded life and had little insight into men's natures; she was naive, and Siraj was the kind of person who could go all out to entice girls. He wondered if he should warn her; but what could he say? She might think him merely jealous and possessive. Perhaps, he thought, it was better not to speak about it at all, but to watch Siraj.

He wished, with all his heart, that Siraj had not appeared on the scene. They had been happy together, and now doubts assailed him. He started wondering about Zohra. She looked cheerful enough with him, but he could not help noticing that she was much livelier in Siraj's company.

The two families met often. They sometimes went to the cinema together, when Siraj almost invariably managed to get a seat next to Zohra and to make whispered comments in her ear.

One night Bashir and Zohra were dining with Shafqat, Jehan Ara, and Siraj. Their cousin Jaffar, who had just arrived in Mussoorie, joined them making a party of six altogether. Siraj was blatantly attentive to Zohra. When drinks were served, he pressed her to try one. But she refused firmly, whereupon he started to sing gaily:

> *Drink some wine,*
> *You will grow divine,*
> *It will work like ancient magic.*

Siraj was irrepressible. Zohra cast a nervous glance at her husband and felt he was furtively watching her. Jehan Ara, noticing her frustration, exclaimed to her brother: 'Oh, stop it!' Then turning to Bashir: 'Siraj is full of such songs.'

The ever-recurring topic of shikar was being discussed at the other end of the room. 'Shikar is the only thing that makes life worthwhile in the districts,' said Shafqat, sounding somewhat pompous as was frequently his way. 'When we are touring we always have friends arranging shikar parties. You must come and see us some time. I promise you shikar to your heart's content.'

'I should like it very much. We might do that during a holiday. It would be a good opportunity too, of seeing you in your native haunts,' said Bashir with a stress on 'native', because of Shafqat's affected mannerisms. He turned to Zohra, but she said nothing.

Jehan Ara exclaimed: 'Yes, Zohra, do come! It's such fun watching. I've sometimes tried using the gun myself, and I assure you it's exciting, even though I've almost always missed.' Her large eyes wandered round the company, seeking applause, but her husband was the only one loyally to comment: 'Yes, Jehan Ara is wonderful!'

'Now, when will you visit us?' she asked.

Zohra, whose spirit revolted at such senseless destruction of life, could only say: 'Apa, I'll come to see you.'

'Oh, Zohra, you are so timid!' Jehan Ara burst into a hearty laugh, her eyes flitting from one member of the company to another. Jaffar was faintly responsive, but Bashir behaved as if he were a wooden dummy.

Jehan Ara had told her husband after their very first meeting: 'He may be all right as your friend, but, poor Zohra, what sort of a lover for a young girl to have! He is hardly human!'

'Were he to respond a little more to your flirtatious ways you would soon think him wonderful,' Shafqat had said, well aware of his wife's weakness.

But Jehan Ara had now become genuinely fond of Zohra, and it was on her account that she wanted to invite them and was even willing to suffer Bashir.

After dinner, Jehan Ara suggested they make up a foursome at bridge, herself offering to stay out with Siraj. Jaffar, standing beside her, was lighting the cigarette she held between her lips as she waited for a reply.

Zohra said: 'I'll be the one to stay out. I have never played bridge in my life before.'

'Then let's drop the idea,' declared Jehan Ara with a puff at her cigarette.

'Don't,' interrupted Siraj, 'I'll entertain Zohra with the gramophone. She'd enjoy that.' Turning to Zohra: 'We have a large selection of records. There are some very good ones.'

'Yes, Apa,' said Zohra, not wanting to spoil their evening, 'You play bridge. I would like to listen to your records.'

The four others moved towards the bridge table on the adjoining verandah. Siraj, turning to Zohra, said: 'I'll select the records.' He wound up the gramophone and started to play some jazz music. Standing and leaning against the cabinet, he took on a consciously studied pose. But soon his whole being seemed to be moving, and he said: 'Your figure is made for dancing. Come, Zohra, I'll teach you.' He held out his hand, but as she did not extend hers, he asked somewhat resentfully with a glance towards the verandah, 'Why? Will your ... will Bhai Bashir mind?'

'I don't now, but I don't want to myself. Play something different,' she said.

He selected 'Drink to Me Only with Thine Eyes', all the time trying to look into Zohra's eyes, but finding them quite deliberately averted. As the record came to an end, he said: 'Let's go out.' As his hand pointed towards the garden, the ring on his second finger glittered almost as if aware of its wearer's mood. 'It's stuffy in here.'

The music and its effect on Siraj were unnerving Zohra. To get away from it, she replied without thinking: 'Very well.'

As they left, Bashir watched them in uneasy silence.

Siraj led Zohra to the hedge over which they could look down upon the hillside. It was lovely, with lights from the distant houses shimmering like giant glow-worms. The moon cast her spell over the scene. Zohra stood watching in rapt silence. Siraj, standing close behind her, inhaled the fragrance of the flowers in her hair. It kindled his already aroused senses.

'You are ravishing tonight! I have been unable to take my eyes off you!' His voice was tremulous with wine and love.

Zohra laughed uncertainly. Impulsively, he slipped his arm round her, and an electric current seemed to pass through her. Alarmed, she wrenched herself away with a quick twist of her body.

'Don't!' she cried indignantly.

He stepped back and for a moment did not reply; then, trying to regain control, he said: 'Is it a sin to fall in love? Does it sound shocking?'

'You know I am married!' Her voice was now quiet, restrained.

After a pause, he repeated, as if still dwelling on the word: 'married!' He shrugged his shoulders. 'What a righteous word for a most unrighteous plan! Marriage should know but one law—the law of nature, the law of love. We would be much better married according to that law. We are both young and mutually attracted. What more do you want?'

'You know you don't really mean it .... Perhaps you are a little upset tonight.' There was a quiver in her voice as her wide eyes looked far away.

'I am not!' he protested vehemently, 'I am only acknowledging the law of nature.' Then, imploringly, 'Let's run away, it would be glorious!'

'Don't joke about such things,' she tried to make her voice sound light-hearted although her lips trembled.

'Whoever said I was joking? Don't I look a wounded man?' he asked mockingly, placing his hand on his heart. After a pause, he continued: 'There is at least one good thing about these arranged marriages. There is no unnecessary waste of emotion.' Taking a step forward again: 'Zohra, don't you ever feel that you should have been allowed to choose for yourself?' Although the question brought back poignant memories, Zohra's instinct urged her to defend her parents and her husband. Therefore, parrot-like, she repeated all the arguments in favour of such marriages, declaring at the end, 'Besides, he is good, and I love him!' She was surprised at her own boldness. One did strange things when placed in an insidious situation. He gave a start at her words, then laughed and moved back a step.

'I don't mean that kind of love. Oh, yes, I know you love him!' He could not keep the sarcasm out of his voice. Then in a coaxing tone: 'Tell me, do you love him as youth should love youth? Does he stir your imagination? Do his words make your heart dance wildly? Is there a symphony of music in your soul every time he appears?' Carried away by his feelings he threw his arms about melodramatically. 'In short, had your parents not chosen him for you, do you think you would have selected him yourself? .... Look into my eyes and answer me honestly!' Moving forward again, he caught hold of her hands. She snatched them away and stepped back.

'What right have you to ask such questions?' Her voice sounded strange even to herself.

'Why are you so angry? Do you think you can fool me?'

Zohra felt confused, distressed. She had not even the power to run away.

'Zohra, my heart, we love each other.' He spoke tenderly. 'Let us go away and be happy together. There is a force drawing us together that is far beyond our power of control. Can you deny this, in spite of your loyalty to Bhai Bashir?' His head was bent in an attitude of pleading.

Zohra was touched by his earnestness, but began to realize, to her consternation, that Siraj had an attraction for her, which she had to resist. She did not know what to say or do. From a life of utter seclusion, she had been thrust into a completely new and bewildering world. Her inexperience and naivety had made her behave with less propriety than she should have, and she had doubtless encouraged him by her foolish behaviour. Zohra felt mortified, but what really frightened her was the realization that it had not been unpleasant.

'It is no use,' she whispered. 'Oh please, don't disturb my life,' her voice now trembled. 'I never quite knew what to make of you—whether you were serious, flirtatious or just fashionable.'

'I confess I wasn't serious at first,' he affirmed in a tone louder than was necessary. 'I thought it would be fun flirting with you. You were so different from the other girls I knew, that you intrigued me. Has anyone told you how utterly fascinating your eyes are? Passionate, yet spiritual—and your smile! I was a fool to think I could play with so exceptional a person and remain unaffected!' Zohra listened quietly. He moved nearer to her again.

'Zohra, think over what I have said,' he pleaded. 'There is no sin in what I'm suggesting. In the eyes of God, we shall still possess integrity. Then, when he divorces you, we'll get married.'

He tried to hold her hand, but she immediately thrust it behind her, and startled by the boldness of his proposal, pleaded: 'No, no! Please forget all this.'

'But why,' he asked impatiently, 'why such a waste? Even our religion, if you believe in it, gives us the right of divorce and remarriage. For a wrong committed by others, surely *you* can't go on suffering a life sentence?'

Zohra's mind wandered, first to her parents and then to her husband, who had shown nothing but devotion to her.

'It is impossible. I respect and love my husband,' she said abruptly and turned away. Siraj was hurt, but his vanity was too great to let him believe her.

'If I were really sure of your love for Bhai Bashir, I'd leave you at once with my blessings for a long and happy married life.'

Zohra, feeling stronger in her resolve, moved towards the fence and leaned lightly upon the railing, silhouetted against the moonlit garden. Siraj lit a cigarette and said wistfully: 'These silver rays make you celestial, while I ... I only become more moonstruck, a lunatic.'

Zohra was feeling calmer now. 'Let us go in,' she said.

'Yes, otherwise they might think we have already eloped,' he retorted bitterly.

9

Nearly a week passed. They had hardly had occasion to be together by themselves again. But Siraj eagerly snatched at moments when, with his enticing smile and ardent eyes, he implored her to yield. He was young and his words carried passion, and Zohra could not help being affected; yet she knew instinctively that she was not really in love with him. But she did gain some insight into love, for in this attraction she realized what love might be. To help her fight this, she involuntarily felt drawn to her husband's protection—the only stable force in this chaos of emotions. It was comforting to feel his strong arms around her. Within them she felt safe, secure, untroubled. She did not know how much Siraj recognized this duality in her nature. One side beckoned her to adventure and excitement, the other, anchored in Bashir, drew her back to safety. The former was superficial compared with this deeper urge to do what was right, whose outer symbol was constancy. It was like a flickering candle-flame compared with the steady glow of the lamplight.

Siraj was restless, seeking an opportunity to meet Zohra. In desperation, he managed at last to arrange a picnic, hoping to gain a few moments with her on her own. But on the morning of the picnic, Zohra felt unwell and Bashir, secretly relieved, sent word that they could not join the party.

Jehan Ara, who felt a sisterly, though somewhat patronizing interest in Zohra, was concerned. On their return from the picnic she suggested an immediate visit to Zohra. Her husband offered to accompany her as he wished to see Bashir. But Siraj stayed at home. He had no desire to call on them, when in all probability he would be unable to see Zohra; besides, even if he could, what use was it meeting her in a crowd again?

Zohra was still in bed when Jehan Ara and her husband arrived. Bashir came out to greet them. Jehan Ara enquired after Zohra and, quite at home by now, walked straight in to see her.

Zohra was lying in a big carved walnut four-poster bed looking pale and depressed. As Jehan Ara approached, she sat up, drawing her legs in. Jehan Ara sat on the edge of the bed.

'Perhaps I overate last night, but I have never been like this before,' said Zohra miserably. 'It is such a shame staying in bed here, in Mussoorie!'

Jehan Ara's queries became more searching, after which a flicker of a smile appeared on her lips as she said: 'Are you sure it's only overeating?' She laughed aloud, casting a significant glance with her luscious eyes. Zohra flushed deeply as the meaning dawned on her when she suddenly remembered her sister. But Mehrunnissa had not been like this, she thought. Not knowing what to do, she drew her legs further in and rested her head on her knees, to hide the strange emotions welling up inside her.

Seeing Zohra's utter unpreparedness, Jehan Ara at once felt sorry and, patting her bent head affectionately, said: 'Maybe it's nothing. I hope it's nothing. I hate girls getting caught unawares, but nature has a grudge against women!'

Zohra was too confused to say much. It was bewildering, and this feeling of nausea prevented her from collecting her thoughts together. She made Jehan Ara promise not to breathe a word of her suspicions to anyone.

When Jehan Ara came out of Zohra's room, she saw Bashir and Shafqat on the verandah sipping drinks. But Bashir instantly got up and came over to her: 'Can you suggest a doctor in case Zohra does not feel better?' he asked anxiously. 'You know almost everyone here.'

Jehan Ara smiled. 'You talk as innocently as your wife! I don't think it's a case for the doctor yet. Let nature take its course.' She sounded pleased with her own superior knowledge. But Bashir looked alarmed and concerned, and Jehan Ara's laughter only grated on his ears.

Bashir had no knowledge of what Jehan Ara had told Zohra, and unwilling to upset her, he did not mention anything himself, but only watched her anxiously. In the afternoon she felt better and was able to get up and dress. And to some extent his fears were dispelled.

But the sickness continued during the next two days. On the third morning, she was worse, and Bashir insisted on calling in a doctor,

whilst Zohra, the suspicions in her heart confirmed, tried to stop him.
'But this is ridiculous! We *must* summon a doctor. What is your objection?' he asked, sitting beside her, as she lay looking tiny in the vast four-poster bed.

'Wait for just another day. I might be all right by tomorrow.' She was trying to delude herself. Bashir noticed her strained face, from which all serenity had fled. He knew she could not but suspect she was pregnant. The possibility of it was being impressed upon him more strongly every day. He had not cared to frighten her until he himself was certain of it, but now, as he sat there, holding her hand in one of his own and lightly running his fingers through her wavy hair with the other, he bent down and in a low voice asked:

'Zohra, do you suspect anything ...?' His keen eyes watched her intently. Unable to meet his gaze, she lowered her eyes.

'I am afraid you are in for it!' His voice was controlled, but his touch was infinitely soothing. 'I wish I could help you. But there seems to be nothing I can do.' He experienced a profound tenderness for this diminutive figure lying before him in the enormous bed, so helpless. All the feverish passion fled for the moment and he was overcome by an overwhelming desire to comfort her. He caressed her gently. Zohra was somewhat calmed.

Under the doctor's directions, Bashir chalked out a regular routine for Zohra. He was confident it would make her feel well again, and therefore enforced it on her even against her inclinations.

As Zohra felt somewhat better in the evenings, he invited Shafqat's family to dinner one day. He was trying to fight that twinge of jealousy for Siraj, as unworthy of himself. He even hoped that Siraj's company would brighten up Zohra, as it had done before.

Zohra was anxious to avoid Siraj, and yet longed to see him.

When they arrived, Siraj noticed Zohra's tired face, and anxiously enquired after her health, but Zohra only became self-conscious and replied incoherently.

At dinner she could not eat much and Bashir, unmindful of the guests, went on pressing her. Zohra, to whom her condition was a new and perplexing experience, could not understand her husband's insistence. It irritated her. But she suppressed her disinclination for food, and forced herself to eat, rather than draw more attention to her condition, until feeling giddy and sick, she had to leave the table. This scene in the presence of her guests, especially Siraj, who she knew was watching her, increased her resentment against her husband's strict regimen.

When the guests arrived home, Shafqat retired immediately, but Jehan Ara, noticing Siraj's preoccupied air, lingered on. Throwing himself down on a settee, Siraj suddenly burst out: 'Why all this mystery about Zohra's illness? Why couldn't you tell me she was in a delicate condition?' The last two words were uttered in a tone of mock propriety.

'But it was none of your business!' retorted his sister. Drawing a chair close to him and sitting down, she continued: 'Besides, Zohra did not wish me to tell anyone. She is newly married and is sensitive about it.'

'Since when have you become so conscientious about other people's secrets?' He looked up sarcastically at her, well aware of her weakness for broadcasting everything.

'There is no secret in this! I didn't tell you merely because you were capable of going and teasing her!' Jehan Ara gave herself away.

'As if I would! I don't see the joke. That old man she has married,' and he swore under his breath, 'I suppose he is quite thrilled. I should feel ashamed. Really, it's preposterous!' His now high-pitched voice was sharp as he turned to her.

'What has all this got to do with you? Surely you couldn't have been serious about her? Anyway, what did you expect?' Jehan Ara's slightly protruding eyes were now bulging with amazement. She was deriving a vicarious excitement from all this.

'Oh, nothing! I'm all right! Only I feel sorry for her. Surely with all his scientific talk he should have had better sense. And the way he was after her at the meal, as if she were a child with no mind of her own.' He shrugged his shoulders. 'Couldn't he see that she wanted to be left alone? Even an owl has more sense!' Siraj's handsome face looked almost ugly as he scowled.

'What is the use of all this abuse? Are these the fumes of a burning heart? Better go to bed with a cold compress,' she said, trying to make light of her brother's outburst.

'If our society is so keen on keeping up its so-called high moral standards, it had better seclude young married women instead of secluding unmarried girls. That would at least save it from a lot of complications.' Producing a showy gold cigarette case from his pocket, he lit himself a cigarette, then continued: 'Girls should be free until marriage, and choose husbands for themselves; then the jealous husbands may put them in purdah, and spare themselves and others unnecessary heartaches.' His tone seemed to suggest that the burden of framing rules for society were best left to him.

'You marry a girl brought up in freedom and try putting her in purdah. She would settle down as docilely as a tigress straight from the jungle, who has been caged!' countered Jehan Ara with a lofty gesture.

'Time yet for that. Anyway, I'm not so possessive!' He was trying to look calm, for his conceit was too great to let him betray his regret. He was already sorry for this exhibition before his sister.

'I am leaving for Lahore tomorrow,' he announced nonchalantly.

'Perhaps that's best for all concerned. We too have only a few days left.' Then she fumbled about for a moment or two and said: 'I can't find the matchbox.' Siraj rose to his feet and exclaimed: 'I hate women smoking!' Nevertheless, he lit her cigarette. He went back to his seat, carelessly threw his half-burnt stub into the ashtray, and lit himself a new one.

Jehan Ara, after a long puff, turned to him again: 'If only you would pay more attention to your studies! You know how worried Father is about your future.'

'He needn't be, and to hell with exams!' He made a gesture as if he were dismissing unpleasant memories. He rose abruptly and, going to the window, turned round to Jehan Ara again. Leaning against the wall, he said: 'I'm joining a film company right now. I think that would best suit my temperament.'

'What nonsense!' protested Jehan Ara, rising from her chair with indignation. 'People of our class don't go in for such vulgar professions.'

'Our kind of people do nothing original,' observed Siraj with bitter sarcasm. 'There's nothing vulgar about art. And acting, to my mind, is the greatest of the arts. Our film industry needs more recruits from a better class of people. If Father expects me to follow in his footsteps and take up a government job, he's in for a terrific disappointment.' Siraj sounded melodramatic. Already he might have been imagining himself a star.

Jehan Ara looked at him with still greater astonishment, for, with her husband in the civil service, government service was to her the height of human ambition. But Siraj had been hearing heated arguments among students on the subject. Gandhiji's Satyagraha movement had affected them profoundly, and the more thoughtful ones were against joining government service and helping the British maintain their Raj. At the time Siraj had only listened to them with limited interest, considering them foolish not to take advantage of the opportunities offered by government posts. But in his present mood, with little taste for studies and a great desire for personal freedom, he seemed suddenly to have changed his mind.

'But this is preposterous! What an idea, going into films of all things!' burst out Jehan Ara again.

'It's the only profession in which one might fully express oneself,' retorted Siraj stubbornly.

Jehan Ara continued to look indignant. But not knowing how to argue with her brother, she changed the subject, leaving it to others to convince him of his folly. 'Will you be seeing Zohra before you leave?' she asked, trying to look friendly and understanding.

'What for?' he almost shouted at her, impatient of what he considered her stupidity. Then, after a moment's silence, he calmed down and said yearningly: 'We might have made handsome screen lovers. But this ... it must be stifling for her.' After a slight pause he added with ironical laughter: 'I might leave a note for her with my advice and blessings. Her precious husband seems incapable of taking care of her.'

The brother and sister stayed up in that strange comradeship till the early hours of the morning, for Jehan Ara with her warm-heartedness was now really sorry for Siraj.

The next day Siraj left for Lahore, and he passed out of Zohra's life as suddenly as he had entered it. Zohra was too sick to feel any great emotion.

She was glad in a way he had made no effort to see her alone before his departure, for in her present condition, she had not the courage to face him. Despite the stern routine enjoined on her by Bashir, she felt no better. Once, as she sat up at night, reading in bed, he came in and, seeing her thus occupied, said: 'Children's bedtime hour!' Zohra paid no heed. He waited for a few minutes, putting his things in order, then came over and quietly closed her book. 'You should not overstrain yourself. If you don't keep regular hours, you will feel exhausted,' he remarked in his professional manner.

Zohra, who was feeling increasingly resentful of her husband's interference in all her doings, could contain herself no longer. 'What do you mean by treating me like this?' she demanded with flaming eyes and a rebellious toss of her head. 'I have borne it long enough. I was hoping that it would make me feel better, but, if anything, I am feeling worse. You leave nothing to inclination!' Her voice choked, as tears rushed into her eyes. Clutching her hands, she continued, for she had been building up her resentment so long, that she had to get it out of her system now: 'It is easy for you to draw up a timetable for me, as if I were a schoolgirl. But if you could only realize how difficult it is to follow it. I am not a machine. I have never been fastidious about my food, but now, when I have strange yearnings, you only try to laugh them off as foolish whims. You try to enforce on me your own kind of diet, which only makes me more sick. In the daytime, my head is so heavy that I can't even read. It is only in the evenings and at nights that I can do anything at all. But you come along and order me to bed. Sleep never comes so early and I lie tossing around for hours. I never knew one could be so utterly miserable. There is not a single hour that I can have relief from this awful feeling. And it all seems so terribly long, so endless ...' She broke down completely and, smothering her face in the pillow, wept from sheer nervous exhaustion. She had never before harangued her husband in this manner, and Bashir had not realized that his care of her was creating such a reservoir of resentment. As she sobbed, her whole body trembling, he bent down and enveloped her with a great tenderness.

'Don't, Zohra, you must not upset yourself! I only thought the right diet and regular hours of sleep would help. I am sorry.' Bashir was greatly perturbed. He tried to turn her round, to wipe away her tears, but she would not let him. At last she regained her composure and, wiping her face, turned round towards him, but without looking at him, said: 'I do not think I should have minded it very much were it not for this awful feeling.' She gave a tired sigh.

'There is surely something that can be done about it. I always thought it would be such a thrilling experience creating the future generation. But this is hopeless. Besides, it is much too early. We have hardly had time to settle down. We were having a glorious time in Mussoorie until this.' Although he spoke quietly, Zohra could see that he was deeply concerned.

After a few moments' silence, she said: 'I wish I was back at home.' For the time being she seemed to have forgotten her home was no longer her mother's house. 'You don't know how much I miss my parents now. So many things have happened to me since I left them.'

As she said this, the vision of handsome, laughing Siraj flashed vividly before her mind's eye, and Bashir, becoming aware of the sudden tension in her body, instinctively guessed the reason. He had hitherto refrained from questioning her about Siraj. But now, holding her head in his hands and pushing back the hair from her temples, he asked: 'Zohra, do you miss Siraj a lot?'

Zohra blurted out all that she could, thankful for the opportunity, for she hated the idea of duplicity. She avoided saying much regarding his feelings for her, for her husband had no rights over Siraj's intimate thoughts. She concluded, 'But all that already belongs to the past.'

Bashir, who had listened without comment, continuing to stroke her hair, said: 'I am glad you told me. I was a little worried.' He gave a sardonic laugh. 'Perhaps I should have been prepared for such eventualities. I once heard a cynic say, "An attractive wife is a liability." But I think she is a liability worth having!' Sliding his hand down to her shoulders and lightly pressing them, he added: 'I have no quarrel with others who also find her attractive.' He was watching her with a keen possessive pleasure. 'Only, Zohra, you are so naive. It would be so easy to deceive you. You know so little about the nature of men.'

His voice betrayed little emotion, as if he were afraid of letting himself be carried away by his feelings. His hands and arms were much more expressive. He was a passionate lover, but he could also be gentle and considerate. She now removed his hand from her body and wove her slender fingers through his, in a token of gratitude for the affectionate way in which he treated her, in spite of everything. But she had not the courage to speak. Bashir, anxious not to cause any more misunderstandings, bent over her: 'You are not angry, Zohra? I only wished to warn you against such people.'

She pressed his hand more warmly and said: 'Now there is no fear. This already makes me feel years older. It was my youth that attracted him.' Looking at him she smiled wearily.

'This will not last for ever,' Bashir reassured her, 'and when it is over, you will still be fascinating and utterly lovely.'

Zohra thought, 'Even if I remained attractive, would that make life any easier?' She realized for the first time that it was perhaps not difficult for her to be desirable to men. Though she did not regret it, she knew she would have to be more careful in her behaviour with them. She recollected how, hardly three months before, she had come to her husband as a bride. She had been tremulous then, with fear that she might not be pleasing to him. She had had no opportunity of recognizing her own allure in men's eyes, and although her husband had succumbed easily, she had not thought that convincing, for after all, he was her husband. But Siraj had been spontaneously drawn towards her and this had made her more wary. She wished she could be a child again, carefree with her parents and her sister.

To Bashir she said, 'Let us go back home to Hyderabad now.'

10

When Bashir and Zohra returned to Hyderabad, discerning and inquisitive eyes instantly gauged the situation. Bashir's mother was, in her own way, exultant, as if it were for her special benefit that the child was to arrive.

'Allah be praised,' she said to all her visitors, 'there will at last be new life in this old home!'

Zohra was mute before such exuberance. However, she soon left for her parents' home.

She saw that in spite of his acceptance, Bashir was unhappy at this parting. He could not conceal the forlorn look in his eyes.

At her parents' house everyone indulged her. No laws or regulations existed for her, except her own fancies and cravings. And, though physically still sick, Zohra improved in spirits. The oppression and gloom that she had at first experienced gradually lifted, and she became more resigned to the trials of pregnancy.

Nalini, too, was now a frequent visitor, and it was she, more than any other, who made Zohra feel reconciled to her condition. Mehrunnissa was spending a week at her mother's with her children. Feeling very knowledgeable and affectionate, she had come immediately on hearing the news and started confiding to Zohra her various experiences, with all the exaggeration and dramatization one might have expected of her.

Unnie and the servants were all overjoyed, but Zohra's parents displayed no such extravagance of spirits.

Unnie noticed this and was puzzled by this attitude. She waited for an opportune moment to speak to her Begum. One day, seeing her seated on the divan in the courtyard, alone with Mehrunnissa's younger son, Anwar, who was just fourteen months old, she came and stood near them. Anwar was seated astride his grandmother's shoulders, with his tiny legs twined around her neck. He was tugging at her plait of hair as if riding a horse, prodding her with his feet and shouting, 'Go, go!' in his baby voice.

The Begum Sahiba, unable to walk around under his weight, was slowly swaying backwards and forwards, to give the child as far as possible the thrill of riding. They presented such a happy picture that Unnie burst out: 'Allah be praised for the grandchildren, Begum; they are the sweetest fruit in all His kingdom!'

The child, on hearing Unnie's voice, quickly slipped down and ran to her, seeking a new diversion. For she was one of those who lavished upon him unrestrained affection and spoilt him unreservedly. Unnie, cracking her knuckles against her temples, handed him her bunch of household keys. She often carried this around, tucked into her sari, with an air of importance.

As the child sat down and set about clinking and separating the keys with his tiny hands, Unnie burst out: 'Allah's miracles are unending! You will have another grandchild yet to fill your lap!' Then in a changed tone: 'But, Begum, to my ignorant mind, far from showing gratitude for this greatest of gifts, you show no great happiness.'

'Allah forbid that I should be ungrateful for His blessings but Unnie, how thin and worried Zohra looks! She is not like Mehrunnissa whom nature has prepared for marriage and motherhood. But Zohra, even as a child, cared for nothing but her books.' Zubaida Begum, stroking Anwar's head, as if to show that grandchildren were always welcome, said: 'I feel sorry for Zohra, especially as she is so sick.'

'*Owi*, a mother's heart is always tender but, graced with motherhood, Chhoti Bibi will soon be well again; she was always fond of children,' said Unnie, not satisfied with the Begum Sahiba's defence.

But Zubaida Begum continued wistfully: 'Allah knows it was I who hastened her into this marriage! I was haunted by the fear of dying before settling my daughter's kismet. Besides, we had set our hearts on the pilgrimage. But I cannot go now, leaving Zohra to the care of others in her present condition. Allah willing, we shall go next year.'

Unnie was about to say something, when the elder brother Shahid ran in and, interrupting her train of thought, clamoured for attention.

II

Zohra had been at her mother's for over three months. Bashir, who now took the liberty of calling at the house oftener, anxiously watched for signs of her wanting to return, sometimes even making mild suggestions. Here, there had to be restrictions placed on their conduct, as any display of affection in the presence of others was considered vulgar and immodest. Besides, it was against his own temperament.

Zohra would never take him to her room. But sometimes the others manoeuvred to leave them alone in the courtyard where there was no real privacy, and a kind of inhibition was imposed upon them. Bashir felt that Zohra was drifting away from him, and he was left with a feeling of desolation. One day, as they sat in the courtyard at a ceremonial distance from each other, Bashir could contain himself no longer.

'Zohra, are you never returning home?' he asked, his eyes upbraiding her.

She gave a start. 'Why, yes. I should, now that I am feeling better.' She was sitting on the *musnud*, distanced from Bashir, for anyone might enter through the open courtyard, although at the moment they all conspired not to intrude. But then one of the servants might be curious.

'You don't look very happy at the prospect!' He laughed in an effort to hide his regret. 'I have had enough of this bachelor existence. I am running after you, trying to snatch brief moments in your company, as though we were courting instead of being respectably married. If you don't come back now, I shall forget altogether what it feels like to be married.' He tried to seem calm, but it was apparent that he was suppressing deep emotions.

Zohra was conscience-stricken. Why had she not thought of it herself? She knew he had been anxious to have her back. Searching his face and looking into his eyes, she asked: 'You don't think I was trying to evade you, do you?' As he made no reply, she continued: 'It is really quite recently that the nausea and sickness have stopped, and you can never know what that means. It was like having a terrible nightmare.' She drew in a deep breath. She so wanted to convey to him that she was neither cold nor heartless.

Bashir now regretted his impatience. She looked so childlike, so penitent for the torture she was causing him. 'Perhaps, Zohra, it is best that you should stay here a little longer. You may not feel as much at home with us as with your own family.' It cost him an effort to say this. He was wondering once more what sort of feelings she entertained for him, although he knew she was not indifferent.

Zohra, watching him, declared with warmth: 'No, no! I will come back. I want to come,' she insisted. Looking at her, Bashir felt happier, for in his present mood any little proof of her love gratified him.

This caused Zohra also to search her heart. She loved Bashir but she felt reluctant to leave her parents and the tranquillity of her life with them. She knew it was not so with her sister, who was always eager to go back to her bridegroom. Was it lack of balance on her sister's part, or was something wanting in her own love for her husband? She could not work out an answer to this. But soon she returned with him to his home.

Bashir was, in his own way, very solicitous. He bought her books on eugenics, and now that she was no longer sick, he insisted on a balanced diet, with proper hours for rest. Every evening he took her out into the fresh air, although this often compelled him to work late into the night and rise early again. Altogether he so enveloped her with care, that Zohra was overwhelmed and willingly submitted to his routine, although in her heart she never relished this clockwork regularity. She was now quite ashamed of her hysterical outburst in Mussoorie. Above everything else, she appreciated the way he shielded her from the possibility of his mother dominating her life. She had feared this most. For her own part she was loyal to her and although many small incidents occurred in Bashir's absence, against which he would have protested had he heard of them, Zohra seldom complained to him, for in her own way, her mother-in-law treated her with kindness. What still jarred on Zohra most was Masuma Begum's ebullience of spirits over the coming child. Besides, it was hardly in keeping with her mother-in-law's personality.

Zohra had as yet seen little of Nawab Shaukat, her husband's aged father. He spent his days in solitude in the men's quarters, cut off from his family. Zohra might not have come to know him, save as a kind of legendary figure to whom she had to pay her formal respects, except he overheard her chanting from the Holy Koran one morning, while he was strolling in the garden. Her voice, full of sweetness, moved him. So, when Zohra, according to custom, went in to offer him her respects, he looked up with a smile and said: 'Daughter, I overheard you chant from the Koran this morning. Allah knows, it was pleasing. If it be no great trouble, come and chant it in my courtyard whenever you feel like it.'

Zohra, in a subdued tone, replied: 'Yes, Abba Jan.'

One evening, Bashir sat down beside Zohra and said: 'I am glad that you sit and read to Abba Jan from time to time. He has had a tragic life and I feel I should tell you of the sadness of this family.' Bashir recounted the story with unaccustomed emotion.

Nawab Shaukat Jung Bahadur had been an influential official in the state government, holding various posts of responsibility, until his retirement at the age of sixty. In his youth, he had been married to Sakina, a gentle, refined and loving wife; the couple had been devoted to each other, but the blessing of children had been denied them. After fifteen years of marriage, when all reasonable hopes were gone, she induced him to marry a second wife.

It was, naturally, not against his own inclinations.

The new bride, Masuma, over twenty years his junior, arrived. Sakina Begum treated her with an elder sister's kindness and affection, but gradually Masuma flung away all submission to such influences, and after the birth of Bashir, her first child, her domination over the elder wife was unrestrained. All Sakina Begum's efforts to share the child and, as was the custom in such cases, treat it as her own were met with cold rebuffs.

This state of conflict lasted for ten years, during which Masuma Begum bore five children, of whom only three survived.

Nawab Shaukat found life increasingly difficult. It became impossible for him to try to balance the scales evenly between his two wives. His religion too permitted him such an alliance only if complete impartiality were possible. It seemed to him a superhuman task.

Thus he was torn between rival loyalties. He still loved his first wife, whilst his second wife's temper he feared; but with the latter was the tie of children.

After the birth of his last child, Safia, as the tension became more and more pronounced, he retired to a life of complete isolation and austerity.

Some years later, Sakina Begum died, which proved a great blow to Nawab Shaukat. For, although he had been living away from her, spiritually he had felt strongly drawn to her. He now relinquished his last official post and became altogether a recluse, absorbing himself in religious and philosophical studies.

'Your chanting of the Koran has moved him deeply.' Bashir took her hand in his.

'Yes,' said Zohra responding to the pressure of his hand, 'gradually a bond of sympathy has been forged between us and I have grown very fond of him.'

Bashir, because his temperament was entirely different, had never been really close to his father, but he was pleased with his wife's understanding of him. In his childhood, Bashir had watched the conflict in their home, without understanding it. He remembered his affection for the beautiful and gentle woman whom he called Badi Amma, Elder Mother. She drew him to her with love and tenderness. He remembered, too, how brusquely the present Begum, whom he called Amma, used to treat everyone, and how he rebelled against her efforts to keep him away from Badi Amma. At the time, he was too young to comprehend the situation. And there were many stormy scenes between mother and child when he insisted on spending more time with the woman who showed so much greater understanding than his own mother. Zohra found that he still spoke of Badi Amma with more feeling than for either of his parents. But now that it was all clear to him, he did not know whom to blame more: Sakina Begum deprived of motherhood herself, desiring to provide her husband with children and arranging this second marriage, or the younger woman married to a middle-aged man for the express purpose of bearing his children, whilst the first continued to be foremost in his affections.

Bashir now realized how intolerable such a position must have been to a proud young woman like his mother. However, he wished she had shown more sensibility and not alienated everyone by her sharp temper and domineering ways.

He often thought that his father was most to blame. He should have foreseen the hopelessness of the situation. But many such instances had proved successful, and from the disposition of his first wife, he could easily have expected a happy solution. For Bashir, the intricate problem had long since lost its edge, but when explaining it all to Zohra, the memories came back to life in a new perspective.

Zohra's heart went out to this frail ascetic man. Her own mother, too, had tried to persuade her father to contract such a marriage and have a son. She now felt thankful that he had not succumbed to her generous-hearted entreaties.

11

Two months before the expected event, Zohra went to stay at her parents' home. The waiting was tedious.

At last, in accordance with Bashir's express request, he was summoned to Nawab Safdar Yar Jung's house in the early hours of one morning, having just gone to bed after working late into the night. He immediately hastened to his father-in-law's abode.

There was no further news; he had to wait. A charpoy was placed for him not far from the Nawab Sahib's, on the open verandah which, during the day, was sheltered from the sun, yet exposed to such breezes as were able to penetrate the sultry heat. But now in the early morning, still dark before sunrise, it was pleasantly cool. He lay down on the charpoy, but neither of them could sleep. The Nawab Sahib kept fidgeting and, from time to time, tried to hold a conversation with Bashir, which always turned more or less into a monologue. He would talk of Zohra and the coming infant and recite fragments of poetry. Bashir, who disliked all effusiveness and was in no mood for talk, was as discouraging as he could be, short of rudeness. But the Nawab Sahib, who at times of stress was inclined to be loquacious, was incorrigible. It was a great relief to Bashir when at last the sky showed flushes of pale pink, heralding the dawn, whilst the stars that had twinkled so gloriously began to fade away. The muezzin's voice came clear and sonorous through the fresh morning air. Everything breathed newness of life, and the Nawab Sahib quickly rose to make ablutions and to offer his morning prayers. Bashir restlessly lingered on his charpoy a little longer. All through the night there had been no message from the zenana.

After the night's vigil, both welcomed the tea sent out to them. There was more cause for impatience now as time dragged on and no progress was reported. Bashir had no access to the confinement room. He was relieved whenever he heard the gentle gurgling of the hookah, for then his father-in-law's voice was hushed, and Bashir could relax a little in silence. His own ashtray was full of cigarette stubs. The Nawab Sahib read only Urdu newspapers, and when they arrived, Bashir eagerly picked one up as a means of distraction. But the Nawab Sahib was too worked up even to attempt such a diversion, and he kept interrupting Bashir. Hurt by what to him appeared disinterestedness on the part of his son-in-law, he spoke with increased excitement:

'Allah knows, before Zohra's birth I prayed fervently for a son, and vowed offerings of rich gifts at the shrines of all the blessed saints in whom I had faith. But Zohra's birth was auspicious. Like a flash she came, and like a flash she entered my heart, and instantly I forgot my disappointment. As soon as I saw her bright eyes, I decided to call her Zohra—Venus—the brightest star in all the heavens.'

Bashir showed more interest in this than he had done hitherto, and the Nawab Sahib, his own appreciation of his daughter mounting in proportion to his agitation, continued exuberantly: 'And since then, I have never once been sorry that she was a girl.'

Bashir, apprehensive of the extended delay in the birth of the child, was growing ever more fearful what this might do to Zohra's already frail health.

At last Bashir asked to see the midwife attending to Zohra.

She came out looking, prim and professional. 'The first labour is often protracted; nothing to worry about,' she said in reply to Bashir's question, as if she were only using an oft-repeated formula.

Still they waited, their anxiety increasing every moment. Eventually, Bashir said, 'We had better fetch a doctor—a specialist; I don't know how efficient this midwife is.'

The Nawab Sahib, walked hurriedly into the zenana, his leisured dignity shed for the moment, and asked to see the Begum Sahiba. She came out of the room and, standing wearily near a pillar, asked with outward calm: 'What is it you want of me?'

'Begum, your son-in-law thinks we should fetch a doctor.'

'*Ai-hai*, Nawab Sahib, what is the matter with you men?' asked the Begum, ready to confront her husband. 'Is a man to enter the confinement room? I have never heard anything more preposterous! Was not this very same midwife successful with our own Mehrunnissa? Has she not skilfully attended to thousands of cases? A little delay, and the Nawab Sahib's senses have gone a-wandering. Shame on you!' Zubaida Begum was shocked beyond measure.

'But our son-in-law has more faith in men doctors,' said the Nawab Sahib, himself doubtful of the wisdom of this idea, but eager now to do almost anything that was suggested. Also he was pleased that, at last, Bashir had shown some real concern, for his controlled behaviour had been puzzling him.

'Allah forbid, this is no work for a man. Your son-in-law behaves like a foreign *sahib* but what has happened to you? And what is beyond my understanding is how a man can know more about such matters.' The Begum Sahiba's tone plainly showed that there could be no two opinions about it. A low, anguished sound from the labour room sent her hurrying back, and the Nawab Sahib, his heart wrung by that cry of pain, went out again, more restless than ever.

The day dragged into afternoon, and now Bashir insisted on having a doctor. After all-round consultations, at which the Nawab Sahib was of little practical help, it was ultimately decided to call in Doctor Ali who would remain outside the delivery room and give an opinion.

The doctor arrived. After discussing the case, he suggested certain injections to speed up the contractions and left, promising to be back soon.

The atmosphere of the house was becoming increasingly tense. Unnie could often be seen hurrying across the courtyard, sharply reprimanding the maids for idling away their time. But there was nothing else for them to do and on this occasion they paid no heed to what she said.

But once, passing by and seeing them all huddled together, Unnie felt the urge to pour her heart out to someone. She moved towards them and, while they waited for one of her usual outbursts, started off, to their amazement, in a confiding manner:

'*Arrey*, Maids, I will cut my hair off as an offering at the saint's shrine when Chhoti Bibi is safely delivered.' Touching her head, she added: '*Ai-hai*, my hair has ripened, but I have never seen such torture before. I start feeling prickly all over whenever I enter that room. I always said that Chhoti Bibi should not be married in haste. But, Allah knows, who was I to impress my views on anyone?' In her eagerness to talk, she did not notice the maids stealthily nudging each other.

'*Arrey*, Maids, I tell you, a girl who has her heart set on books should not be rushed into marriage. It is different with us simple folk,' she summed up and, scratching her head, hustled away.

Hardly was her back turned when the maids burst into smiles and giggles. It was not so long ago they recalled that she had voiced exactly the opposite opinion on the subject. Unnie's sudden change of views brought relief from the tension of Zohra's situation, and Gulab, with hands on hips, addressed the others: '*Arrey*, Maids, the oracle hath spoken; what if it be after the event?' She shook her head wisely in imitation of Unnie. As everyone giggled appreciatively, she burst out in a different tone: 'The old witch! Wait and see if I don't get her head clean shaved by a proper barber after Bibi's child is born!'

But those last words reminded them again of the gravity of the present circumstances. They immediately lowered their voices and started to talk in hushed tones.

Bashir was now accompanying the Nawab Sahib into the zenana quarters quite freely but, while his father-in-law often lingered on to talk more intimately to those around, Bashir stayed for brief moments, merely enquiring after his wife's condition. Seeing the maids and women gathered in little knots, with glum faces and talking in whispers, Bashir was irritated. He could often hear 'Allah, Allah' on their lips. Why could they not do something about it, be more active, he thought. As if Allah had nothing to do but listen to the supplications of such women! But instantly another thought drifted through his mind, silently, unconsciously. But what else could anyone do? What was he himself doing? They, at least, had the strength of their faith, whilst he was adrift, as it were, in a vast mysterious ocean with nothing to cling to.

As he passed Zohra's room he could hear long deep groans. He wiped off the beads of cold sweat collecting on his forehead.

He was exasperated by everything and everybody. Science, God, and Nature, all alike, seemed to fail him.

Once again as the two men waited in the verandah, the Nawab Sahib, who had been smoking his hookah, suddenly pushed it away, and with a look as if he had been inspired by a vision, said gravely: 'Let us consult the poet Hafiz. His prophecies always come true.'

Without waiting for a reply, he rose and briskly walked away. After making his ablutions he came back and picked up a volume from its niche. Sitting down cross-legged, he unwrapped the old brocade in which the precious volume was wrapped, and drawing the Kashmiri silver reading stand in front of him, reverently placed the volume on it. After reciting a verse from the Koran, he opened the book at random. As the leaves fell open, Bashir saw him fix his eyes on a passage and read it carefully to himself.

Bashir watched the proceedings with an air of detachment and lack of enthusiasm, yet with an impatient fascination at the depth of his father-in-law's faith. The Nawab Sahib, having carefully weighed the meaning, now smiled serenely and translated the Persian couplet into sonorous Urdu for Bashir's benefit:

O cup-bearer, light our cup with the radiance of thy sparkling wine
O singer, sing thy song, for the world is moving in the wake of our desires.

The older man's eyes lit up with relief as he savoured each word, and tranquility seemed to descend on him, as it might on a soldier after a hard battle won. 'I have every faith in Hafiz, the greatest Sufi poet of all time. His words are always prophetic,' he observed enthusiastically.

Bashir was moved, in spite of himself. As he sat there, he mused whether it was really possible for him, an oriental—an Indian—to detach himself entirely from the ageless background of his forefathers. In his subconscious mind, did there not still linger a spark of faith or superstition, or whatever else it was called? Was it only buried under a thin veneer of science; would just a little fanning rekindle the embers of that faith? The idea faintly amused him, even though he was in no mood for amusement.

The household was gripped by extreme unease. Doctor Ali having returned, there was a further consultation. Apparently there was no reason for such delay. Zohra's strength was ebbing. She had gone through over sixteen hours of intense pain. Despite her youth, she would be unable to endure the ordeal much longer. They decided on the use of forceps; getting everything ready, the midwife went in to perform her task. The Nawab Sahib sat down to read the Koran, calm and unperturbed, now that Hafiz had spoken. He knew all would come to a successful fruition. Of the two, it was Bashir who was now more nervous and restless. He had little faith in the prophecy of poets, and science seemed to be letting him down. Anxious to be alone, he went out for a stroll under a sky brilliant with stars. But the stars carried no message for him; they were mere reminders of the commencement of another night. As he paced the garden path, his mind went back to that day when he had first entered this house to fetch his bride. His mind wandered, recalling the past year. Although their time together had as yet been so short, he suddenly realized with fear clutching at his heart that it already felt like a distant memory. It was almost eleven months to the day since their marriage. His bride, whom he had never seen before their wedding day but to whom he was instantly attracted, had been unable to speak, too shy to meet his eyes. Her timid but exquisite ways had drawn him to her in a strong protective desire, and he had done all that was humanly possible in the circumstances to allay her nervous fears and to act with restraint. For himself, they had been the happiest months of his life; but what of Zohra? Would she have chosen this life if she had had the power to choose? However much he searched his mind, he could find no satisfactory answer.

At last, seeing his father-in-law coming out, Bashir hastened to him.

'Allah be praised! All is well! You are blessed with a son!' The Nawab Sahib embraced his son-in-law warmly.

'How is she?' asked Bashir anxiously, freeing himself rather abruptly from the embrace.

'They say she is all right, though wholly exhausted,' replied the Nawab Sahib.

They walked in, but could now gain no entry into the zenana, nor was any further news conveyed to them for what seemed an endlessly long time.

At long last they were called into the zenana apartments and shown the infant. Bashir struggled to appear calm, giving him an air almost of indifference, as he gazed upon his son; but the Nawab Sahib's eyes were misty with tenderness and love for his grandchild. The child was wrapped in an old shirt belonging to the Nawab Sahib—this was considered auspicious.

Bashir was able to get a brief glimpse of Zohra as she slept the sleep of exhaustion after nearly twenty hours in labour. Relieved by this peaceful sight, he went home.

Zohra slept on without disturbance till the muezzin's call to morning prayer woke her and she saw the dawn creeping in through her window. When she awoke she instantly asked: 'Where is the baby?' She had not even seen it yet.

'He is with your mother,' said the nurse.

'Fetch him, please, I want to see him!'

The nurse brought the infant in and placed him beside his mother. As Zohra watched her son with a shy worn-out smile, he began to move his small rounded limbs and opened his eyes through narrow slits. Her heart gave a leap, and something of immense intensity rushed into it suddenly. At that moment was born within her an extraordinary and unexpected sweetness, wild, passionate and imbued with ardour. She had borne this child through all those long months merely because there was no escape. But all of a sudden now, her son filled her with a tenderness she had never before experienced. He seemed to be the fulfilment of all her desires. The weary months of sickness and of waiting, the long unending hours of agony, when she had been desperate for any relief, even death, was all in the past and unreal compared with this reality. As he moved his tiny hand, she lifted it up to her lips and kissed it with a tender fervour transcending any that she had ever known. She caressed his smooth soft face, his hands, his feet, and was filled with rapture. It all seemed so incredible. She was happy that she had seen him alone, this first time. In the presence of others she would have been too shy to express her feelings. The nurse, a professional, was taking little or no interest in her reactions. Zohra drew the little infant closer to her, clutching him to her heart, which seemed to burst with happiness.

The baby began to cry and the nurse took him away, leaving the young mother in a state of delirious joy. All her disappointments, a marriage that she did not want, a home which could never be her own, the school that she had been forced to give up, all these sacrifices were now converted into an ecstasy of completeness. Into this overwhelming current of love, her husband was also drawn. All at once she felt closer to him. It seemed that the child bound them together as nothing else could have done.

II

Zohra and her infant were surrounded with all possible care and love. Relatives and friends, mostly women, came constantly to see the new mother and child. Marriages and births hold a fascination for all women; but to those in purdah, they are the pivots around which their entire lives revolve. Gifts were showered on the infant.

Bashir, in those first few days, was distressed and annoyed at constantly having visitors surrounding his wife. One day, coming out of Zohra's room, he found Zubaida Begum alone in the courtyard. He went and sat near her. Although he was wearing a western suit, having come directly from work, and squatting was not easy, he would not be persuaded to take a chair whilst his mother-in-law sat on the carpet. Somehow she always moved him to greater respect and consideration than anyone else he knew. She was busy chopping with a silver nutcracker, almonds, pistachios, and other nuts into tiny bits to prepare the specially nutritious sweetmeat given to young mothers after childbirth. Looking up at her son-in-law, whilst her hands continued to manipulate the nutcracker, she asked: 'Mian Bashir, how do you find your wife and child?'

It was the cue he wanted. Without much ado, he said: 'Mother, there are too many people coming to see Zohra. You must curtail these visits.'

'*Owi*, Son, Zohra is well now. How can she be happy lying alone in that room?' she asked, looking at him with her delicate smile. 'It would be like solitary confinement. She needs laughter and company to recover her health.' The Begum Sahiba spoke freely to Bashir, who always found her charming and gracious.

'But, Mother, so much company is sure to tire her and have a harmful effect on the child. Zohra is so anxious for him.' Bashir was insistent, though respectful.

'*Ai-hai*, how can I turn out our friends? Have sense, Son,' she said in gentle rebuff. 'Besides—may no evil eye fall upon her—are not both Zohra and the child looking well?' she asked, handing him nuts on the tray, knowing that he did not relish paan.

Taking the little tray and thanking her by slightly raising himself on his knees and offering salutations, he said: 'The child is flourishing, but Zohra often looks tired; in the west they would never allow so many visitors.' He put bits of nuts into his mouth with great decorum.

'Allah forbid that we should act that way. I do not profess to understand either foreign ways or foreign people. Anyhow, to me, the *mem* looks frigid.

How can she have a warm heart?' asked his mother-in-law, looking seriously at him.

This only evoked a smile from Bashir. He was amused at her summary way of dismissing members of an alien race. Besides, he thought, it was convenient to dismiss all English women as frigid and at the same time have a craving for their white skin. Surely it was a complex created by fairer skins having been successively in power for centuries and centuries—ever since the Aryan invasion. But he merely said: 'Underneath our different skins, human feelings are essentially the same, it is only the mode of expression that varies.'

'Allah be thanked that here we act on inclinations and impulses,' she said fervently. 'And have you considered what our relatives would say if we were to shut them out in this inhospitable manner merely because Zohra might get tired? No, no, I cannot do that! .... *Ai-hai*, Son, is she not my daughter? Are you the only one to care for her?' she asked in a quietly challenging way accompanied by a conciliatory smile.

Bashir knew it was useless trying to press his point. Besides, his mother-in-law's personality had a strange fascination for him, and he was ready to concede anything more easily to her than to his own mother.

Youth is resilient; and Zohra, in spite of all the strain she had undergone, recovered quickly. The infant was named Mohammed Shameem, but for all practical purposes, only the second name was used.

The fortieth day was celebrated with great feasting and rejoicing, accompanied by alms-giving. Zohra looked radiantly happy in the crimson and gold sari presented to her by her mother for the occasion. She had suddenly acquired a serene dignity, which with her still girlish ways was very engaging. Yet, although her face had regained the warm colouring of her first Mussoorie days, she was not allowed to return to her husband's home.

Bashir, beginning to be fretful of his loneliness, could not understand this delay. His ideas and theirs about such things were so radically different. But it was impossible for him to persuade Zubaida Begum, and he had no alternative but to conform to their ways.

Zohra, also, was content to stay away and this, as always, hurt Bashir more.

It was not until the child was three months old that Zohra returned to her husband's home.

12

Zohra's existence now revolved round that tiny life with a force far beyond any human power to control. A hundred meanings unfolded themselves to her in his simple movements that no one else could fathom; it was in the nature of a revelation that left her spellbound.

Once, while Zohra was seated on the divan, playing with her infant, Bashir entered the room. He stood still for a moment, struck by the look of adoration in her eyes. They seemed to caress the infant. She hardly noticed his presence.

He came and sat down beside her. The sight stirred him to ask: 'If you could have foreseen this, would you have deliberately chosen all those weary months of sickness and of waiting, and that terrible ordeal?'

Zohra gazed into the clear baby eyes, as if in them was written the answer. Then, turning to Bashir, she gave him a smile that expressed more than she could ever have put into words.

Bashir himself had taken a long time to forgive the infant the anxiety of which he had been the unconscious cause. But now he went on, half in jest, half in earnest, probing into Zohra's mind:

'So will you want more children?'

'Why not?' she retorted unhesitatingly, with a light still lingering in her eyes. Then seeing her husband's bushy eyebrows lift in amazement, she added: 'Oh, I did not mean immediately!'

'This is beyond me. I should have thought one experience was enough to put you off for ever.'

'Nothing worthwhile, they say, is achieved without travail and risk. Here, at least, the reward is far greater,' she answered tenderly.

'Zohra, I cannot profess to understand you, but it is good to have you back again.' He kissed her, with his warm lips on her smooth throat below the ear lobe; it tickled her. Laughing and moving away, she lifted up the child to him.

'Kiss my baby!'

Bashir pressed his lips to the child's forehead, saying: 'He is mine also.'

'But he is much more mine!' she exclaimed, her large eyes flashing defiance. She resented his casual manner towards the infant.

'By law, a child belongs first to his father,' he said, teasing her for her insistence.

'What do you mean?' she asked emphatically. 'What have you done for him? You do not even show much interest, whilst I ... he is part of me. You can never take him away from me!'

'But whoever said I wanted to take him away from you?' he asked, unaffected by her outburst.

'But supposing ... supposing something were to happen between us; you would have him then? It is wrong. It is sinful.' Her soft voice now rose in passionate revolt.

'But what do you expect to happen between us?' he asked, not taking her seriously.

'I said "supposing" ... but why should a father have prior right to a child?' she asked, looking genuinely hurt.

'*I* did not make the laws, Zohra.' He tried to laugh away her earnestness. But in her new consciousness of motherhood, she could not take this lightly.

'Men have made laws to suit their own purposes. There is more slavery in this than in anything I can think of. If they wanted to put a woman in bondage, there could have been no heavier chains to bind her down. For the sake of her child a woman would stay with her husband even if she intensely disliked him.' She spoke earnestly, holding out her hand to the infant, who clutched her finger tightly.

'What do you want then?' he asked, to help relieve her anxiety. 'The matriarchal system?' There was amusement in his eyes for, in her gravity, Zohra looked younger than ever.

'No, not altogether.' She was irritated by his imperturbability. 'You can have all the rights after me. I do not like maternal uncles stepping into the picture. Anyway, in this case there aren't any!'

'You can have whatever rights you want,' he countered indulgently, for by no stretch of imagination could he foresee such an eventuality. Then he added, more in his professorial style, 'Zohra, you are too young to spend your entire life so engrossed in your son. Why can you not trust him a little more to the ayah? It would do you good to get away sometimes, instead of sitting and brooding over the infant and getting strange ideas.' Then, expressing what he had felt keenly ever since her return: 'You know, you have been neglecting me altogether. What good does it do me to have a child, if I am to lose my wife?'

'But this is only until he is a little older,' she said, feeling guilty.

'Yes, and then you want more children!' His keen eyes looked ironically into hers.

'Still time for that,' she gave a hesitant smile.

'In spite of all your fierce ideas on motherhood, you do not look any older than the girl I married,' he observed, placing his hand affectionately on her shoulder. 'You have acquired the possessiveness of the old, while still retaining your naive ways and the combination makes you appear even more youthful.'

'Call it possessiveness or what you will. But I shall fight these man-made laws some day!' Her words no longer had any force in them. She laughed happily, although with a determined toss of her head.

II

Zohra's life flowed steadily like a quiet refreshing stream. Shameem proved not only the chief source of pleasure to herself, but also to her mother-in-law. She showed such a capacity for love that Zohra wondered what she might have been like had fate blessed her with a more normal life. Within that stern exterior, what hopes lay buried, what passions lay smouldering, she herself would not know. Zohra could never quite understand why she had ever been given in marriage as a second wife to an elderly man. Unlike her own mother, Bashir's mother possessed a disciplinary mind, and as Shameem grew older, Zohra relied more and more on her judgement. The child's existence proved a great comfort to Masuma Begum in every way, for there was a new fear now gnawing at her heart. Her second son Hamid's return from England was long overdue. He had taken a Tripos in

Modern Languages at Cambridge University. But he had not yet said a word about returning home.

Bashir, who, on account of his father's self-imposed seclusion, managed all their affairs, was exasperated. Once, coming straight from his mother, he said to Zohra: 'Hamid has no sense of responsibility, no consideration for his parents. What on earth can he still be doing there? But, although he is no fool, one should not expect wisdom from him either.' Bashir's voice was cold.

'What is the matter with Bhai Hamid?' asked Zohra, distressed by her husband's tone.

'Matter? He alone knows what the matter is with him. He seems to have got it into his head that the whole world is his special concern. I should not wonder if he were trying to smuggle himself into Russia, to get first-hand knowledge, as he used to say, of what is happening there. Or perhaps some other similar folly!'

Zohra, realizing the subject was upsetting to her otherwise level-headed husband, did not pursue it. But she observed that her mother-in-law and Safia adopted quite a different manner in speaking of Hamid. Masuma Begum once confided: 'Dulhan—she would always be called 'Bride' by her mother-in-law—'there are strange misgivings in my heart. Why does Hamid not return? *Owi*, Allah forbid, but what shall I do if he brings a *mem* with him?' There was a pathetic note in that harsh voice, as she looked around as if already wondering how to fit a foreign woman into these surroundings.

Safia, on the other hand, declared in her unthinking, carefree way quite oblivious of the agitation she was causing her mother: '*Ai-hai*, why all this fuss? I only wish he would return, even if it be with a *mem* as wife. What does it matter? Let him marry whom he pleases.' She already visualized a separate establishment for him, where she herself could be a frequent visitor. Then impulsively: 'Hamid Bhai Jan is very lovable. You know, Bhabi Jan, once I thought you two ...'

She suddenly collected herself and broke off, making the pause still more awkward, as Zohra understood the unspoken thought.

When Shameem was two years old, Zohra's second child, a daughter, was born to her at her mother's home. It was an easy confinement. The child was named Shahedah.

Zohra returned to her husband's home after the prescribed rest. Afterwards she looked back upon this as the most peaceful and contented period of her married life. She recollected Sarojini Naidu's line:

Born in my life's unclouded morn.

13

*M*eanwhile, there was no news of Hamid. It was nearly a year since Bashir had stopped his allowance. Masuma Begum was getting more and more despondent, when suddenly they received a wireless message from the ship, *S.S. Franconia*: 'Arriving Fifteenth Bombay Love Hamid.'

Bashir showed no emotion, but his mother and Safia were overjoyed.

Immediately on getting the news, Safia exclaimed: 'Amma Jan, I am going to Bombay to receive him!' She flung her arms out with excitement, and began to plead with her mother, who was sitting on the divan with Zohra.

'But, how can you go alone? Will your bridegroom accompany you?' asked her mother, pleased but cautious.

'I'll ask him if he can get leave. But if he can't, surely Bhai Jan can come!' Safia spoke aggressively, as if already there had been a refusal.

'He might, but he too will have to apply for leave,' countered Zohra, loyally rallying to the defence of her husband.

'Oh, I'm sure he could if he tried!' said Safia, flippantly shrugging her shoulders.

Although Masuma Begum disapproved of what she considered her daughter's disrespect, she was keen that Bashir should go to welcome his brother. She herself could not even think of it, as staying in a hotel was, to her, like stepping into another world. Knowing Zohra's influence over her husband, she turned to her: 'Dulhan, why don't you both go?' Her voice sounded more like a command, but Zohra was now used to it.

'Yes. Also, it would be a good opportunity of seeing Bombay and doing some shopping,' added Safia.

'But, Safia Apa, what about the children?' asked Zohra.

Safia herself did not quite recognize the complexity of the feelings that she had for her sister-in-law. Zohra was attractive, intelligent, and accomplished—loved by all. Safia was none of these things. Since the birth of her children, Zohra's beauty had taken on a special quality, and serenity and contentment seemed to engulf her whenever Shameem and Shahedah were with her. The deep void in Safia's own life, created by the lack of children, often made her tense and irritable in Zohra's presence. Now she interrupted impatiently: 'Leave your children to Mother.'

'Yes, Bride, may Allah preserve them. Can you not entrust them to me?' asked Masuma Begum.

Zohra, abashed, said: 'Amma Jan, it is only that I have never left them before.'

'Then, Bride, it is time you learnt to do without them for a while,' rejoined her mother-in-law, ready to take control.

Afterwards, when Zohra spoke to Bashir about it, he said, contrary to all expectations: 'If I can get leave, it would be excellent. We shall have a holiday from the children too.' Although he spoke without apparent irony, Zohra knew it was a criticism of what he considered to be an obsession with her.

Bashir, Safia, and Zohra left for Bombay four days prior to Hamid's arrival. They wanted to look round the city. Bombay was not a centre of ancient culture, but among its main attractions were the shops dealing in foreign goods. Recently however, there had sprung up the Swadeshi Market, a narrow lane, with small booth-like shops on either side displaying Indian wares. Walking through this, Safia squeezed Zohra's hand and ecstatically exclaimed:

'*Arrey*, Bhabi Jan, these lovely, lovely saris! I feel like buying them all!' Zohra, looking around dazzled by the beauty of the fabrics, returned the pressure of her hand. But Bashir said drily:

'I don't know how your husband tolerates your extravagance, Safia.'

Safia turned away with a mocking shrug. Zohra knew that Safia, fond as she was of dressing up, was also desperately trying to live up to her dashing husband's standards of stylishness, and her heart went out to her sister-in-law. Bashir, she felt, should have displayed greater sensibility. But Bashir was unaware he had given offence. To appease Safia, Zohra stopped in front of a shop where saris were hanging down like flaming banners, and said:

'Safia Apa, let's buy some of these khaddar—hand-spun—saris. I have been longing to wear them.' She sounded enthusiastic.

'Nonsense, Zohra, none of that for me. I am going to buy these fine mill-made ones,' she said, turning round to the shop opposite.

Zohra was sorry that her sister-in-law should express contempt for khaddar. It had come to signify the fight for freedom since Mahatma Gandhi's advocacy of the use of hand-spun cloth, as a double-edged weapon—political and economic. Bombay was the cradle of this new life. Khaddar-clad women and white Gandhi-capped men, symbols of a newly aroused nationalism, were to be seen everywhere.

'I'll buy these, they fascinate me,' said Zohra, stroking the rough khaddar saris with her hands. 'They look so pure. I feel as if perhaps they have a soul.' She spoke hesitantly, fearing that neither her husband nor Safia would understand.

'Well, whatever you do, please hurry up. Once you women start shopping, there's no end to it!'

Seeing Bashir's impatience, Zohra pleaded:

'Do you not also want to buy something? This lane is full of fascinating things.'

Before he could reply, Safia burst out impatiently: 'But why do you have to come along with us? We can manage by ourselves, and you can occupy yourself in some worthier manner.'

'Use some sense,' said Bashir, looking disapprovingly at Safia. Her manners also jarred on him. 'How can you two wander around all alone?'

'Why not? This isn't Hyderabad that our dignity should suffer! Besides, nobody here is going to take any notice of us. You talk of our being helpless, and when we want to be independent, you treat us with contempt.' Safia shrugged her shoulders, and turned back to the saris.

Zohra, in her heart, agreed with Safia. But she disliked the confrontational stance that brother and sister adopted towards each other in such arguments. Anyway, like a rebellious martyr Bashir stuck to what he considered to be his duty to the women, and dampened their spirits.

There were better cinema houses in Bombay than in Hyderabad. Besides, they could all sit together instead of the two women being segregated in the zenana section. So, on two evenings they went to the cinema. Safia also made use of every possible moment Bashir happened to be absent to gossip with Zohra. The chief topic was Hamid, and as she talked incessantly about him, Zohra once laughingly said: 'The number of times his name is mentioned, your Hamid Bhai Jan can scarcely be having breathing space from his hiccoughs.' Nevertheless, she herself became somewhat infected with her sister-in-law's enthusiasm. Sometimes, in a more confidential mood, Safia would tell Zohra how much more fun they could have had if Yusuf were with them. But he was always so busy. Zohra listened to such outbursts sympathetically; but herself observed absolute silence, which Safia, in her eagerness to talk, hardly noticed.

At last, the long-awaited day arrived. Early in the morning, they all went with garlands of jasmine and roses to receive Hamid. As they watched the ship come slowly alongside the wharf, Safia was the first to detect a figure standing alone at one end of the deck, leaning over the railings on his elbows. With a joyous exclamation to her companions, she waved to him. He immediately responded by waving his handkerchief. Bashir, despite his cool composure, was somewhat moved at the sight of his only brother, and almost involuntarily waved back in welcome. But Zohra, although excited at the idea of meeting her brother-in-law, remained motionless, watching him keenly. She wondered if Hamid was unsociable, for she saw his fellow-passengers walk up to him, and apparently for lack of encouragement, move quickly away to form other groups. Hamid remained aloof, framed against the sweeping mass of humanity on the moving ship. She saw him often pass his fingers through his hair; sometimes she thought he was pulling at it. There was a pipe in his mouth, which he handled constantly.

'Mother's fears were groundless. Hamid Bhai Jan seems to be all alone!' exclaimed Safia, nudging Zohra.

'I am glad, otherwise she would have been greatly disheartened,' said Zohra, with a sympathetic feeling for her mother-in-law.

At last the gangway was in place and visitors were allowed to go aboard. 'You take one garland and I'll carry the other,' Safia said to Bashir, as she picked up hers from the basket.

'You can garland him for me also, since you ordered mine. I don't see the sense of making a hero of him after the way he has behaved.'

Safia was annoyed, but having no time to argue, quickly thrust the garland into Zohra's hands and moved forward eagerly, saying: 'As Bhai Jan is unwilling to welcome his brother back, even in this most ordinary manner, Bhabi Jan, will you please do it?' She cast back a hurried pleading glance.

Zohra turned to her husband who was following close behind, wondering what she should do.

'I do not mind, Zohra, if you garland him. Only, I wish to be left out of all this.'

They had already started going up the gangway. Zohra felt her husband was in the wrong; and she was hesitant at the prospect of garlanding a stranger, even though he was her husband's brother.

As they reached the deck, they discovered that Hamid had travelled deck class. This further infuriated Bashir; to him it was dragging down the family name.

As they moved towards Hamid he hastened to meet them. His glowing face showed no sign that he was suffering from any sense of lowered dignity. Zohra, stepping back, watched Safia greet him as she flung the garland round her brother's neck. Brother and sister met in an affectionate embrace, Safia's eyes filling with happy tears. Hamid, too, was obviously moved. Bashir, trying to disguise his feelings, shook hands, and made an effort to return his brother's embrace. Turning to Zohra, Safia, as if she were presenting her special discovery, exclaimed with obvious delight: 'Hamid Bhai Jan, meet our Bhabi Jan.' Safia keenly watched his reaction.

They took a step towards each other. Zohra was about to offer salutations in the Hyderabadi fashion, but seeing him extend his hand, she put out hers; it was not a warm clasp, but the quick nervousness of a sensitive touch.

Safia reminded her whispering: '*Ai-hai,* Bhabi Jan—the flowers.'

With an embarrassed laugh, Zohra flung the garland round Hamid's neck as he bent slightly to receive it. Then, with a jerk, he straightened himself. Their eyes met and held each other for a brief moment. There was a certain aloofness in his look, which she could not define although she felt it had a quality of dreaminess intermingled with a hint of sadness. Yet they were friendly eyes with almost a look of recognition. Zohra noticed he was several inches taller than she was, but stooped a little. He was informally clad in a beige sports shirt, an indescribable blue-black tweed jacket with leather elbow pads and grey flannel trousers, all rather worn out. Even so, he seemed to have a style about him that immediately attracted attention.

The day passed quickly. In the afternoon, Safia and Hamid were left to themselves. Seated in a corner of the lounge, ordering endless cups of coffee, and Hamid puffing at his pipe, they talked for a long while. Hamid showed an eager interest in all news concerning his parents, Safia's husband, their relations, and friends. But Safia in turn was able to get little out of him concerning his life in Europe.

After dinner, they all sat on the balcony of the Taj Mahal Hotel, overlooking the harbour. Even after the long separation, the brothers had little to say to each other. It was Hamid and Safia who mainly kept the conversation going. But Zohra examined him stealthily through half-veiled eyes. His light-brown complexion was pale. His head nearer to the classic Greek mould than anyone she had yet seen. His wavy black hair, a bit too long, was all disarrayed, what with his fingers forever straying through it. His long-lashed eyes were lively, protected by frank well-defined eyebrows that jutted out slightly; furrows were already beginning to appear between them. The medium-sized nose was well set above lips, which were apt readily to burst into a smile, through which flashed white even teeth. The sweeping outline of the jaw, with almost an eager chin, gave the final touch to a sensitive, slightly long oval face. It presented a fascinating study to Zohra. She also noticed that, except for the too frequent use of English words, he spoke Urdu fluently, although his pronunciation betrayed a slight foreign accent. Often students, on their return from abroad, behaved as if they had forgotten their mother-tongue. In contrast, Hamid's spontaneity in immediately switching back to it was so refreshing, she thought. This was more marked as, all around, Indians even amongst themselves, conversed in English, though in a variety of accents. But Bombay was an astonishing city and its elite were more westernized than their brethren elsewhere in India.

For fear of being discovered in her scrutiny of her brother-in-law, Zohra often diverted her glance to the sea, where the moonlight was dancing on the waves in streaks of quivering silver. Hamid, whose own eyes often turned that way, silently observed her.

Zohra looked very lovely that evening in a golden-coloured sari. She still retained the slender elegance that belonged to her father's family. But two children had made her figure slightly fuller, which only added to her allure.

Her face, with dark enticing eyes and dimpled smile, glowed with warmth. Her curly hair was tied up in a loose knot at the back, encircled in a chaplet of heavily scented *champak* flowers. Around her long and shapely neck hung a single gold chain. Hamid's aesthetic sense delighted in her beauty as he watched her graceful movements.

Safia, having talked about a great deal of other things, again asked him about himself, but the only reply was: 'There's really little worth telling. Most of it would only bore you.'

But Safia, no longer able to restrain herself, said: 'We thought, perhaps, some romance was keeping you there.'

He gave a cynical laugh.

'There's no longer any romance in Europe.' He tapped the ashtray with his pipe and bent his head as if in reflection. 'The war killed all such luxury of emotions. There is only reality left.'

Safia gave him a quizzical look.

'Oh, I don't mean I had become a sadhu! I was not aspiring to such spiritual altitudes. There was no reason why I shouldn't meet girls or like them.' He knitted his eyebrows, which immediately gave his expression the appearance of being closed—withdrawn.

But Safia pursued relentlessly: 'You mean you had flirtations or what they call love affairs?' she asked, unable to restrain her curiosity.

'Since when have you taken upon yourself the duties of a father confessor? I'm not seeking absolution yet.' In a gesture of dismissal with the pipe in his hand, he indicated his desire to bring the subject to a close.

Safia gave Zohra a sidelong glance, but her head was turned towards the sea. She wished Safia would stop this intrusive questioning. Bashir sat in apparent unconcern, glancing through the newspapers lying before him on the table.

'Anyway, you can tell me what else you were doing,' persisted Safia.

'I travelled a bit, finding odd jobs and staying in strange places. It was exciting.' He again made a movement in the air with his pipe.

As Hamid turned to Zohra, she realized he was trying to get away from his sister's inquisition.

'Do people in Europe like Indians?' Zohra asked.

'On the Continent, people are very friendly,' he said, smiling at her, as if in gratitude for rescuing him from Safia. 'They show great interest in our ancient philosophy. Actually, it is due to Mahatma Gandhi; for wherever one goes, people are anxious to know something about him. His non-violent movement has caught their imagination. And no wonder, after the soul-destroying war!' Then, trying to include his brother in the conversation, he turned to Bashir and said: 'Science certainly has made Europe a cleaner and healthier place to live in. The standard of living is so much higher.' He shrugged his shoulders and, with an impatient gesture, returned his pipe to his mouth.

'You will admit that science has made the people of Europe both physically and mentally more fit,' said Bashir. 'Isn't it amazing how even the poorer class of children look healthy and happy there? Their life span is always increasing.'

'As for that,' said Hamid, 'I call it prolonging the agony of existence.'

Zohra gave a start; she observed that he now had a kind of tortured look.

'There must be some more drastic method of revolutionizing society,' he continued. 'We in India need it more than any other country.' He spoke with fervour and his eyes glowed.

'Revolution!' Bashir, aware of his brother's socialist leanings was perturbed. 'Anarchy and bloodshed! Murder of decent people, destruction of everything worthwhile! Is this where you lovers of India want to lead her to?' There was frozen contempt in his voice. 'Besides, the Islamic social order is the best. There is no class distinction; no accumulation of wealth. I really don't think you can improve upon it.' Although he hardly practised the precepts of Islam, Bashir was a staunch supporter of the Muslim social structure.

'Our religion may not encourage class distinction, but what about our society? We need all kinds of radical changes. But our first and most imperative duty is to get rid of foreign rule, and for that Mahatma Gandhi's way is the *only* way.' He spoke with intensity, striking his hand on the table. Zohra noticed that his fingers were long. His voice was full of feeling, in direct contrast to his brother's.

Bashir retained his equanimity. He said: 'I have lost faith in Gandhiji's methods since he suspended the mass Civil Disobedience movement in 1922. At that time, England's embarrassments were many—Ireland, the Middle East. Had Gandhiji provided unflinching leadership then, we may well have gained our independence by now. Now we can only move in a constitutional manner, and swaraj will come gradually. The time for revolution is past.'

'We cannot wait for anything so gradual. We cannot let the masses continue to live in semi-starvation. We must strike a blow at once against the system that makes all this possible.' Hamid was getting worked up.

'I agree that the standard of living of our masses must be raised. But with our ever-increasing population, this is an enormous task. No power can handle it easily.' Bashir spoke calmly as if delivering a lecture.

'In Europe, these are days of high speed for everything but the birth-rate, whilst in India everything is at low speed except that.' Hamid's derisive laughter was infectious. Safia and Zohra found themselves joining in, but Bashir's irritation at his brother's ideas and ideals was manifest. He said he was tired and went up to his room.

The atmosphere immediately became more relaxed, although Zohra felt a little guilty at her own feeling of relief.

'Then, where were you for so long?' Safia reverted to her old subject, which she had found difficult to pursue in the presence of her elder brother. 'Paris? I hear it's very romantic over there.'

'Really Safia, are you insinuating that there's nothing else in Paris but women?'

'But what else is there?' asked Safia with a mischievous gleam.

'Yes, Paris is gay, Paris is intoxicating,' admitted Hamid, 'but above all, Paris is the heart of creative genius. Parisians are artists in every aspect of life. Theirs is a living culture. But in a way, I was more fascinated by Hungary and Rumania. The peasants are noble and friendly, and the loveliest handicrafts still survive there.'

'*Ai-hai*, Hamid Bhai Jan, I like machine-made things best, they have such finish,' said Safia, secure in the superiority of her own taste.

'To me handmade objects have a beauty of their own, a certain dignity, a distinction,' he said enthusiastically.

'Why, that was just the way Bhabi Jan talked when she wanted to buy khaddar saris. Oh, you Gandhian disciples!' exclaimed Safia, turning from one to the other.

As Hamid looked at Zohra, a glance of understanding passed between them.

Hamid now talked more freely, recounting incidents, which amused Safia and Zohra. At last, Zohra thought she should leave brother and sister alone together.

'I think I shall retire now,' she said, rising.

Safia made no effort to detain her, but Hamid remarked, 'Must you go?'

'Yes,' she answered, 'besides, we have bought toys for the children. I must pack them carefully.' Her eyes lit up as she mentioned the children.

'I am looking forward to seeing my little nephews and nieces.' There was a questioning note in Hamid's remark.

'Only one of each,' she explained smilingly, and moved away.

Hamid watched her receding form. 'Bhai Jan has found a very attractive wife,' he observed.

'You remember my teasing you about Zohra?'

'Yes, it was when you pronounced her name this morning that I suddenly recollected. Somehow in your letters I didn't recognize it. But anyway, I am glad she is married to brother. The very idea of the permanency of marriage fills me with horror. I hope you've no one else on your hands to dispose of!' he said half-seriously.

'Why shouldn't you marry if you find someone suitable?' she asked, with a coquettish toss of the head, which with her was not very becoming.

'Who's to be the judge—you?' he laughed. 'Anyhow, Brother seems to be more in love with his wife than I should have believed of him.'

'Yes, in his own way, he indulges her in everything.' Hamid could detect a tinge of envy. 'If you or I say things he disagrees with, you know how intolerant he is; but if she says the same things, he pampers her as if she were still a child.'

'And she?' he asked, getting interested in his brother's family.

'Oh, she appreciates his devotion and all that! But you can't expect her to get romantic about Brother, can you?' she asked, admitting to him what she would not easily have admitted to another.

'I don't know, why not? I suppose much would depend on her temperament,' he replied, wishing to be fair to his brother. 'Judging from first impressions, I should think she'd be a soothing influence on any man. In these days of neurosis and hectic living, that in itself must be like a tonic, though Brother hardly suffers from nerves.'

'Only you and I seem to ruffle him up,' she said happily. For in her present mood, even a weakness shared with Hamid made her feel fashionable as he had just returned from abroad.

But Hamid musingly continued: 'In that golden-yellow sari, she somehow reminds me of a daffodil. There is something in her so refreshingly genuine.

Her manners and gestures are perfection, and she moves so gracefully. There is really nothing discordant about her. Even her voice is soft.' He smiled as if he had attempted to find some flaw, but had failed.

'*Arrey*, Bhai Jan, she seems to have captivated your imagination,' said Safia, proud of her choice of bride for her brother.

'Yes, aesthetically she pleases me,' he admitted. 'But she looks too young to be talking about her children. What are they like?'

'Shameem is not quite three-and-a-half. He is getting very headstrong and unmanageable. Mother says he's just what Bhai Jan used to be. Shahedah is a little over a year old, and is adorable. She has salt in her face.'

'I had forgotten that phrase. The English would say "sweet". It's typical of the differences in our tastes. They love chocolates and sweets while we prefer something spicy.'

'Their food is tasteless,' said Safia. 'I've only tasted it in these hotels, and I can eat nothing without pickles.'

'But altogether, we've cultivated a taste for over-spiced things,' said Hamid quietly. Seeing Safia perplexed, he merely added, 'Anyway, it will be nice to have the children in the house.'

'You love children and you hate the idea of getting married. It doesn't sound logical!' exclaimed Safia, with a hint of exasperation.

'One must have some reason for wishing to perpetuate oneself. I'm afraid I don't deserve that distinction. Of what use would it be to have more people like me who cannot harmonize themselves with their surroundings? I'm all out of tune, a kind of broken symphony. Bhai Jan is the right person to leave his stamp on posterity.'

'Our family is dependent only on Bhai Jan to carry on the line. It's good that Bhabi Jan is fond of children. I think I'll adopt the next one.' She tried to laugh, but her voice became harsh, and Hamid discerned real anguish.

'But will they, will she, consent to give it to you? Anyway, why should she choose to have any more?' he asked seriously.

'*Arrey*, yes, that she is sure to have,' exclaimed Safia confidently, 'and according to our practice, I have an unwritten right to one of Bhai Jan's children,' stated Safia, as if already determined to exercise that right.

'Poor Zohra, she has to provide children for the whole family! Some responsibility! In that case I might also apply for her seventh or eighth child.' He wanted to bring the talk down to a lighter vein.

'She is young enough to have a dozen,' laughed back Safia, with a carelessness that could only spring from not having been through the tedious process herself.

'How old is she?' asked Hamid.

'Twenty-three.'

'She looks younger. Somehow, one cannot associate her with children,' said Hamid as he became lost in thought.

14

Whilst Hamid played with his brother's children in the zenana room overlooking the courtyard, his mother watched them from the divan. It was a sight that melted her stern face into tenderness.

Zohra strolled in from her side of the house and, without mounting the divan, sat down on the edge. Hamid, lifting up little Shahedah, came over to them.

'Amma Jan, please make me another paan. I feel as if I had never been away. Nothing seems to have altered except, of course, that the clan has expanded.' He gave Zohra a smile, as if welcoming her into the fold.

His mother said: 'Allah knows how much worry you caused me, Hamid. Seven years. I counted the days, the weeks, the new moons, and grew weary. Allah be praised a thousand times that you are back and without a *mem*!' There was indescribable relief in her voice.

'And what if I had married? Would you have disowned me, Amma Jan? There are nice girls everywhere.' It amused Hamid to think of the way the two races confronted each other. There were difficulties, no doubt, owing to their different modes of life, but the gulf appeared even wider because of their status as ruler and the ruled, and more pronounced by their different skin colours.

He heard his mother sigh as she said: '*Ai-hai*, Son, this is no matter for jesting.' Carefully preparing a paan, she was handing it to him, when Shahedah in his arms, snatched it away. Whereupon Hamid, closing his eyes, coaxed her to put it into his mouth. This she did with great glee. He kissed her tiny fingers, as she rippled with laughter. Both the Begum Sahiba and Zohra watched them happily.

'Kiss your uncle, Shahedah!' exclaimed her grandmother. Shahedah, putting her arms round him, kissed him delightedly.

The Begum Sahiba, turning to Zohra, said: 'Bride, look how quickly Hamid has made friends with his brother's children!'

'Yes, Amma Jan. It is amazing the way Shahedah clings to him.' Her heart swelled with maternal pride; but there was also some wistfulness, for her husband never showed such genuine enjoyment in their company.

'*Ai-hai*, Bride, in the excitement of Hamid's homecoming, I forgot to see your new saris. Fetch them here. I want to see the new Bombay fashions.'

Her voice had its habitual authoritative note, but Zohra no longer minded it. She rose to comply with her wish. Returning with a bundle, she placed it on the divan, and sitting down beside it, picked up the folded saris one by one and displayed them before her mother-in-law. Hamid, immediately interested, drew closer, and stood by the divan. As Zohra held one up to the sunlight filtering through, he remarked: 'There is nothing to rival our sari for grace and simplicity ... when rightly worn.' He almost unwittingly threw an appreciative glance at his sister-in-law, who, becoming self-conscious, pretended not to have noticed it.

But instantly, she heard her mother-in-law say: '*Owi*, Allah forbid, Dulhan, are you going to wear this white sari?' She sounded greatly displeased.

'But, Amma Jan, it has a coloured border; and so many girls and young matrons wear white these days,' said Zohra, nervously biting her finger tips.

'*Ai-hai*, you girls have all gone crazy, Allah forbid that I should allow my daughter-in-law to commit such folly! Bride, I thought you had more sense.' Her voice, pained and reproachful, broke off. Zohra was abashed, and listened in silence. Hamid, noticing her confusion, felt the urge to defend her.

'But, Amma Jan, why cannot Zohra wear it? I am sure it would look striking.'

'This is not a matter for you, Hamid,' she replied curtly, as if closing an unpleasant subject, 'and I do not like the modern way of addressing your elder brother's wife by her name.'

'But, Amma Jan, what else shall I call her—Mrs Bashir?' he asked teasingly, trying to restore her to good humour.

'Yes, anything that savours of the foreign! I suppose it is beneath your dignity to call her Bhabi Jan?' she asked sharply.

'She is years younger than I am. It is absurd my trying to act the younger brother to her, unless she wishes me to ... Do you, Bhabi Jan?' he asked with mock gravity, looking down at her. She turned her head towards him, trying to suppress her amusement, but neither pursued the subject.

Later in the afternoon, they strolled into the courtyard and to the fountain in the centre. Zohra sat down on the edge, beside Shameem, who started throwing crumbs to the goldfish. Hamid, standing a little aside, lifted up Shahedah, so she could watch the scene. As the children became absorbed in their pastime, Shahedah clapping her hands each time a crumb was swallowed, Hamid turned to his sister-in-law: 'Zohra ... er ... Bhabi Jan ...'

'Don't,' she interrupted, 'I feel foolish when you address me so respectfully. Remember, I too should then have to live up to the role of elder sister.' With an arm around Shameem she was looking up at him, a smile flickering on her lips. She looked so young that Hamid could not help smiling back at the mere idea.

'I am glad you are not keen on such formalities,' he said. 'One can surely be better friends without them.'

Shahedah, with a child's possessiveness for a newly acquired friend, could not bear to see Hamid's attention diverted. Placing her fingers on her uncle's mouth, she lisped: 'Chacha, Chacha,' and pointed to the fish again.

Zohra, watching them, said: 'Shahedah seems to think you are her private property! I hope she does not make a nuisance of herself.'

'She gives me back my lost youth,' he said with a boyish enthusiasm that made his remark sound ridiculous.

After a while, the ayah came to take Shameem and Shahedah away to the maidan where the children of the neighbourhood gathered to play. Shahedah accompanied her reluctantly, after a great deal of persuasion from everyone.

The ayahs were often more eager to go than the children themselves, for it gave them the opportunity to gossip about their masters and mistresses. Intimate knowledge of high society seemed to bestow upon them an elevated social status. Nowadays they were full of Hamid's homecoming.

After seeing the children off, Zohra sat down again on the edge of the fountain, while Hamid remained standing.

'Why did Amma Jan get so upset over those saris?' He came straight to the point, for his mother's behaviour had rather surprised him.

'Our elders still consider white, for women, as essentially the colour for widows. I had hoped, however, the bright green border would be sufficient to atone for the inauspicious association; besides, this was sort of opalescent,' she said, dipping her hand into the water, and hardly looking up at him.

'Poor Amma Jan! It is the unhappiness of her own life that makes her hold on so firmly to such beliefs,' he said, and Zohra was touched by the compassion in his voice.

'Most elders cling to superstitions, like we all do to heirlooms,' she observed. 'The connections have too long been part of their lives to be easily jettisoned. I am sure that my own mother would feel the same.'

Hamid appreciated her loyalty in defending his mother's unyielding position.

'I can never understand our attitude towards death.' He looked perplexed. 'According to our faith, death should lead us—at least the good amongst us—to Paradise, the land of plenty and eternal bliss. Amongst the Hindus, it leads to a higher stage in spiritual evolution. As such, we should not have such a horror of it.'

'Our elders take a morbid delight in talking about their own deaths,' explained Zohra. 'It is only when the word is mentioned, however obliquely, in connection with their children, that they find it so frightful.'

'What I cannot understand is how those who are intensely unhappy themselves can go on praying for long lives for their children. They should at least condition it by saying: "If they are happy."' As he spoke, Zohra noticed a darkening expression flit across his eyes.

'That would be wiser,' she admitted. 'But, after all, the urge for parenthood is rooted in the continuity of life. Happy or unhappy, it must be sad to see one's children pass away before one. It is the frustration of the very purpose one is expected to serve.' She was thinking of her own children.

'Still, we must try to develop a different attitude towards death. It would ease the parting too.' Hamid moved closer to the fountain and stood looking into the water, when Zohra said:

'Two years ago, my parents went on the pilgrimage to Mecca and our old Unnie also accompanied them. They each brought back for themselves white coffin-sheets, dipped in the sacred waters of the Zam Zam spring which is reputed to have sprung from the spot where the prophet Ismail's feet struck the ground. I teasingly asked Unnie if she had brought one for me too. You should have seen her expression of horror. From the look on her face one might have thought that it was an eventuality that could *never* arise.'

'Were youth to die, it would appear they would deny it sacred burial,' he commented.

'Does death fascinate you?' she asked, looking up at him.

'Yes, anything unknown is intriguing; and death, after all, is the greatest mystery; to project oneself into eternity; to lift up the veil and glimpse beyond the barrier.' His eyes seemed to be trying to penetrate the limits of his understanding.

'I like the way Rabindranath Tagore invokes it: *O thou the last fulfilment of life, Death,*' she quoted.

'Are you fond of poetry?' he asked.

'Yes.' Her eyes lit up. 'Are you?' she asked with an inclination of her head.

'Yes,' he replied, just as briefly; but his eyes spoke much more. 'But now, I should like to read more of Persian poetry, especially the Sufi mystic poets.'

'You should meet my father,' said Zohra. 'He is keen on it. He often entertains poets and literary men. Sometimes he arranges *mushaeras*. It used to be fun on nights when these poetry reading sessions were arranged. We used to listen to the poets reciting their verses, from behind the purdah.'

'Why "used to"—don't you now?' he asked.

'No, not usually, unless I am staying with my parents.'

'Is your father's home far away?'

'Not far from Golconda Fort. But your brother does not care for *mushaeras.*' Then as if trying to defend him, 'Besides, he really has no time.'

Somehow an unconscious restraint seemed to have sprung up in their conversation.

With an effort she proceeded: 'Father has some rare manuscripts of the Mughal times. He loves to show them to appreciative visitors. Calligraphy, as you know, was an art in those days and some of the illustrations are like gems. Perhaps you would like to see them some day.' She talked on, seeing that he was interested.

'I should love to,' he said.

Zohra now got up and, wandering around the courtyard, started to pluck jasmine blossoms. Hamid, putting away his pipe, helped her. Inhaling deeply, he exclaimed: 'They are heavenly!'

'I shall go in and string them,' said Zohra.

'I am going to Safia's. Why don't you also come?'

'It is time for your brother's return. He expects me to be in,' she said with a reticent smile that always came to her when she spoke of her husband.

Hamid merely said: 'It must be pleasant having a wife to welcome one after a tiring day.' He wanted to say—'an attractive wife', but refrained from making such an intimate comment.

15

Hamid was reluctant to discuss his plans for the future with the family, for whenever the topic came up, they could think only of the various government services. But he could not respond to their suggestions with any zeal, and they began to fear he would go away again and take up work outside Hyderabad.

One evening, when Safia was present, they all assembled in the courtyard where it was cool. Masuma Begum was seated on the divan with her paan box in front of her, looking imperious, as her eyes proudly wandered over her brood. To her it was like a miniature court, with Safia and Zohra sitting on either side of her; Hamid, on a cane chair close to them, was describing Venice with its canals and palazzos and the haunting songs of the gondoliers, while Shameem and Shahedah, running backwards and forwards from the fountain, often distracted their attention. They all looked up happily when Bashir returned from work. Seeing this gathering of the clan, he pulled a chair for himself and sat down. Tea was ordered for him.

As they sat talking, the children paid little attention to their father, but tugged at Hamid's hand attempting to drag him away from the group. Hamid gently tried to put them off, but they were insistent. Bashir turned to Zohra and asked: 'Why have the children not gone for their walk yet?' There was mild censure in his voice at the indulgence shown them.

Before Zohra could reply, the ayah, who was standing near the fountain, came forward.

But Bashir, still addressing his wife, said: 'Send them away!' His voice was firm, but not unkind. Zohra silently rose and went over to the children. Shahedah was clinging to Hamid's knees. Zohra with an apologetic smile tried to free her. Hamid stood up to help her, and between them they persuaded the children to go. Hamid carried Shahedah to the gate and there handed her to the ayah. When he came back and resumed his seat, an uncomfortable silence descended upon them all.

Safia, resentful of her elder brother's disciplinary measures and lack of tenderness, turned to Hamid saying:

'Hamid, Bhai Jan, please, oh, please, don't ever go away again! The family does not feel like a family when you are not here.' Safia, half-reclining, leaned against the end of the big round bolster at her mother's back.

'But whoever said I was going away? Why should I?' Hamid felt uneasy at his own importance in the family circle.

'Allah be praised a thousand million times for this!' his mother was moved to say. 'But, Son, have you decided yet what you will do?'

'In a way, yes, Amma Jan. I was just thinking it over before telling you, when Safia ...'

'Yes, thank Safia for broaching the subject,' burst in Safia. 'Amma Jan, sweeten my mouth, for Hamid Bhai Jan is staying here.'

'Do you not think we should at least know what he intends doing?' asked Bashir, always sceptical of his brother's good sense.

'What does it matter so long as he stays in Hyderabad?' exclaimed Safia carelessly, playfully sifting through her fingers the tiny bits of chopped betel-nut in the bowl in front of her mother.

'Hamid, what service are you thinking of joining?' his mother asked.

'Amma Jan, why do people here only value government service? To me, it is like being placed in a sacred niche,' he said, his hand going to the collar of his knee-length, tightly buttoned-up sherwani as if he were trying to free himself from something stifling. Zohra had noticed that he repeated this gesture every time he felt unsure of the response his remarks would provoke.

'Then what will you do?' asked his mother, looking at him with astonishment.

'I was thinking of starting a bookshop,' he stated.

'Allah forbid! You have such strange ideas, Hamid. *Ai-hai*, are you going to work in a shop?' his mother looked aghast. 'Your grandfather of sainted memory never served anybody. He only looked after his estates, and led the leisured life befitting his status. He was generous to the poor, and you should have seen the way they worshipped him. By the grace of Allah, he had an elephant in his stables, which he mounted on ceremonial occasions. But times and outlooks change, and your father,' here her voice hardened, 'joined government service on a salary; and now the son wants to serve behind a counter, become a shopkeeper! *Ai-hai*, what is the world coming to!'

Her pleasure at Hamid's staying in Hyderabad was now transformed into consternation.

Bashir intervened: 'You presumably want a shop for foreign books. I do not see much scope in that.' He sounded as if he would finally dispose of the subject. The only effect it had on Hamid was to make him more firm in his resolve.

'It would fill a great void in our city's life,' he countered. 'There is not a single shop here where one can get a fairly good selection of modern books.'

'You mean you want to popularize leftist literature.' Bashir's voice was constrained but underlying it was a note of intolerance.

'I think it is imperative that people should know of the great modern experiments taking place elsewhere in the world. We must shake ourselves out of this inertia; free ourselves from mental bondage to a decadent system. Above all, let everyone have the freedom to think and act for himself. We should not tolerate any system in which there is not complete personal liberty.' Hamid's voice rose with mounting passion. 'I love Hyderabad and value much of its old-time charm. But our ancient feudal order, however benevolent it might be in some ways, must change if we are to value human dignity more.' His eyes were burning. His hand struck emphatically against the divan.

'In other words, you want to undermine our youth with your impractical ideas, and help create chaos and disorder. You talk of equality, socialism, and communism, and would at the same time love to have beautiful things around you. You can't stand this, you can't stand that; it does not please your aesthetic sense!' Bashir was caustic.

'It is just because I value beauty so highly, that I would like to make life beautiful for others too,' said Hamid. 'I would like them all to have an opportunity to cultivate the aesthetic sense. We want to raise the living standards of the poor, not to lower ours to their level. There should be some common meeting ground, a halfway house, between the two extremes— luxury and poverty.' Hamid was speaking from the depth of his heart, in a voice charged with sincerity.

'And in the monotony of sameness, kill all initiative,' observed Bashir drily.

'No,' said Hamid firmly, 'and it is because of it that I cannot subscribe to any of the prevalent *isms*. Regimentation of spirit is intolerable. With Gandhiji we shall have to evolve a way of our own.'

The argument was becoming heated. Masuma Begum, though emotionally always on Hamid's side, was mentally now on Bashir's. And no one else dared to speak, so they went on arguing. Safia, meanwhile, wearied by the turn the conversation had taken, had quietly slipped away.

Hamid at last said with exasperated finality: 'Whatever happens, our sort of life cannot go on for ever, and the sooner we realize it the better!'

Before Bashir could reply, Zohra, who was now pouring out the tea for him and had listened intently to the argument, turned to her husband: 'But have you not yourself said that life here was stagnating and needed a fresh impetus?' She upbraided him mildly, as leaning forward she handed him his tea.

This seemed to have a calming effect on Hamid, but Bashir, accepting the cup, started impatiently, 'Yes, but ...'

'But,' she interrupted, straightening up and ignoring her husband's cold anger, 'but only your methods differ from your brother's, and why should he not make his own life?' She was unconsciously quoting his own words. Bashir resented this defence of Hamid, but it made him realize he should control his temper.

An awkward silence ensued, broken only by the sound of their mother chopping betel nut for the paan.

Bashir, though trying to restrain himself, was unwilling to change the subject as that might be taken for acceptance of defeat. He said: 'But supposing you enter on this venture, do you really think there is enough scope in it?' He was now calm and tried to sound detached.

'There are all sorts of other possibilities,' answered Hamid. 'We may also have a publishing house of our own. Young writers are frustrated everywhere in India for want of encouragement. I am convinced that some such venture is essential. We can also start a reading room and then, gradually, circulating libraries.' Zohra found it easy to be infected with his enthusiasm, but Masuma Begum said in a tired voice:

'I expected you to join the political department like your father. There would be bright prospects there.'

'Amma Jan, I am afraid I am not made for political glory; I can never shine in such a firmament,' he remarked pleasantly, now that they were all restored, outwardly at least, to good humour. Then, as if admitting a failure, he added: 'In a way, brother is right. I am always in conflict with myself. Emotionally, I still like much of our cultural heritage, but intellectually I feel we have to struggle for greater equality. I know some of the old-world charm, the gracious way of life, may be lost in this, but we have to risk it. We cannot have starvation and deprivation in our country, and still be happy.' Hamid's sensitive face was charged with feeling.

II

In the days that followed, Hamid was diligently working out plans for his bookshop but still found time to visit his old haunts.

One evening after dinner, Bashir retired to work whilst Hamid and Zohra remained with his mother. Suddenly Hamid announced: 'I feel like going round and seeing the city by night. I wonder if it looks at all changed—the lights and all.' He knew his mother would never stir out for anything she considered so flighty; therefore, turning to his sister-in-law, he asked: 'Will you come, Zohra?'

'I shall go and see how long your brother still has to work.' She went in and returned swiftly. 'Yes, I will come.'

Masuma Begum looked on with a smile of indulgence. She remembered Hamid's solitary broodings before he went abroad, and wished to safeguard him against getting into that mood again. The society of a young sister-in-law would do him good, she thought.

Hamid drove his parents' car with Zohra beside him

'Let us first drive down to the Char Minar, it is after all the heart of the city,' said Hamid.

'And the centre of gossip too!' Zohra exclaimed. Then quietly: 'But gossip does often revolve round the heart, does it not?' She looked up, questioning, but found his eyes were fixed steadily on the road.

'Well, it should not here, with girls safely protected behind purdah,' he said, taking a turn to the right.

Crossing the bridge, they passed through the old city gate and reached the Char Minar. This simple impressive square structure stood in the middle of the city. It was said that in 1591 the kingdom of Golconda was besieged by plague for which no cure could be found. The ruler Quli Qutb Shah in desperation had a *taziya* placed in the centre of what was to be his new city, imploring Allah to have pity on his suffering subjects. When the pestilence ended, he built this triumphal arch in the shape of a *taziya* with a dome and arches and the four minarets, which gave it its name, as a thanksgiving to Allah for having shown such mercy. Large earthen water-pitchers placed by the fountain in the centre invited the thirsty to quench their thirst. From the Char Minar, the city of Hyderabad spread in all four directions.

Some beggars, homeward-bound, seeing the car slow down, rushed towards it, demanding alms. A man of erect bearing, in long faded orange robes, with unkempt hair hanging down almost in ropes to his shoulders, was the first to attract attention. He carried a black rosary hanging from his wrist and a beggar's bowl in his hand. With one hand on the rosary telling the beads, he cried in a slow sonorous voice: 'In the name of Allah, give alms to the fakir and you and yours will be blessed!'

Hamid searched his calm face, wondering whether that look of inner peace was real or faked.

Zohra was touched by the sight of a woman in filthy rags, with a naked child at her breast, demanding alms. She was uncomfortable that Hamid also should see her thus bared. A pained expression came over Hamid's face as he handed her a coin. The woman cried hysterically: 'May Allah bless you with thousands and thousands of years of life! May you live for ever like the sun and the moon in conjugal bliss, and may you be blessed with armfuls of sun-like sons, to perpetuate your blessed race!'

'What a bombastic blessing!' Hamid burst into a subdued laugh. 'Besides, any man and woman seen together, and they must be taken for husband and wife.'

Zohra offered no comment, for she found it embarrassing to have her brother-in-law mistaken for her husband.

As they turned into one of the bazaar streets, Hamid noticed a flower shop, from the arch of which hung fragrant garlands and bracelets in beautiful clusters.

'These bracelets look lovely. The buds are set like pearls. I think I will buy them for you.' Paying no heed to her protests, he parked the car on one side and got down. He soon returned with a pair of jasmine bracelets.

'They are very pleasing.' Her eyes thanked him as she inhaled their fragrance. Then she wound them round her wrist.

The car slid by gleaming bangle shops. Endless rows of bangles glistened in their glass cases. Hamid, slowing down to almost crawling pace, asked: 'Would you like to buy some bangles?'

'It takes time to select and you would find it boring,' she protested, not wanting to overtax her brother-in-law's patience.

'No, I am enjoying all this.' He looked genuinely interested. 'I remember, before I went away, I sometimes came here with Safia on her shopping expeditions. It is always more pleasant at night, cooler and no need for purdah. I wonder you do not come more often at this time—or do you?'

'No, somehow I do not care to come alone at night, and your brother has no liking for it.' Then collecting herself: 'Besides, he has no time.'

Hamid merely observed: 'The bangle trade seems to flourish here.'

'We Hyderabadi women have a mania for bangles,' she said, 'and happily enough, they are also essential. If I went about without them, Amma Jan would be more horrified than even by that white sari. You know, bare wrists signify widowhood.' She looked at him as if she were not sure whether he was aware of this.

'Anyway, they give a woman the right feminine touch,' he said with an appreciative glance at her hand.

Hamid, who was driving very slowly now, stopped in front of a bangle shop that looked tempting, and a woman came up to the car to enquire what kind of bangles they wanted. Whilst Zohra examined the samples brought to her on a tray, Hamid, seeing a man still working inside the shop, got down to watch. On a lacquered foundation laid over the glass of the bangle he was making an intricate design with sequins and seed-pearls, afterwards sprinkling it with gold dust to cover up the tiny crevices still showing. There was a small coal oven in front of him over which he warmed the lacquer.

Zohra, well aware of men's impatience at shopping, hastily selected a set for herself and two tiny bangles for Shahedah. Hamid, returning to the car with a glance at the workman, said: 'How late these people work! That man is making an exquisite design, but I wonder what he will be paid for it.' His eyes were thoughtful as he leaned against the car, resting his hand on the window.

As the bangle seller approached with her parcel, Zohra asked: 'Do they usually work so late?'

'Yes, Begum Sahiba, quite often all through the night. There is a bridal set to be completed and delivered tomorrow morning to the bride's home. May Allah preserve the rich, but weddings cannot wait for bangle sellers; we have to make sure that the bangles are ready in time for the wedding day,' she replied loquaciously.

Back in the car, as they drove slowly on again, Hamid recited a couplet:

> *Some are like fields of sun-lit corn*
> *Meant for a bride on her bridal-morn*

'*The Bangle Sellers*' by Sarojini Naidu,' retorted Zohra, and continued:

> *Tinkling, luminous, tender and clear,*
> *Like her bridal-laughter and bridal-tear.*

'You should know what that means,' he said with a slight inclination of his head and a quick glance.

'Not the laughter,' she answered with a nervous laugh in which Hamid could discern a certain wistfulness.

'Wasn't Bhai Jan a stranger to you when you married? I imagine it must have been quite frightening for you,' he said, then added hurriedly, 'for you both.'

She held her wrist against her face, deeply inhaling the fragrance from the jasmine bracelets and did not look at him. Noticing her reticence, he did not pursue the subject. The car moved on. In the white latticed balconies that overlooked the streets, Zohra could see the silhouetted forms of women looking down at the passing panorama of life. They had come out onto the open balconies to amuse themselves and to breathe the fresh air.

After passing the city gate and crossing the bridge, Hamid turned down the road running parallel to the riverbank.

The High Court, designed in the Moghul style of architecture, with gleaming white domes and turrets, was lovely and impressive. The hospital on the opposite side was also modelled on similar, though simpler, lines.

Hamid parked the car and they both got out. Standing on the bank of the Musi River, they gazed upon a scene very familiar to them. A half-moon shone in the dark blue sky and the calm shallow waters reflected it, without a ripple, as if lulled to sleep by the serene light.

Buildings, gleaming white in the moonlight, stood stately against a background of the silver grass of rising riverbanks. High above the city walls they could still see the Char Minar. Hamid and Zohra, silently watching the scene, no longer felt they were strangers to each other, but companions drinking from the same fountain of familiar beauty. Hamid broke the spell.

'These half-real visions out of one's childhood are strangely overpowering. They are in no way the world's loveliest sights, yet, they seem to stir one most.' Hamid was still gazing across the river, his hands on the railing.

'Yes, this magic of belonging is bewildering,' said Zohra drawn back to reality. 'Even a dream in which I was compelled to stay away from Hyderabad, I think, would become a nightmare for me.'

'Yes, the love for places does get into one's veins. It is like a lover long parted from his beloved. Away from her, the enchantment dims; face to face again, the fascination returns two-fold. But I should not indulge in it; neither the love for a place, nor for a woman. It is stifling, narrowing and it enslaves you!' He made an impatient gesture.

Zohra wondered if he were speaking from some personal experience. But he looked like a man who wanted no limitations and who, being tied down by fate, was chafing under it like a spirited horse in harness. Only, his was intellectual impatience. His eyes, now in friendly converse, turned full on Zohra, and met hers, deep, dark, and illumined. 'How strange that this girl should be the wife of a brother, with whom I have nothing in common except parentage,' he mused to himself.

Zohra had been watching with interest from underneath her lids, the slim lean figure beside her. To her he looked lonely, almost forlorn.

'Sometimes, I feel I should have got away from all this and worked in wholly different surroundings,' he said, 'but it is kismet. We are irrevocably chained to the twin wheels of life and fate. There is no breaking free from them.'

There was a kind of defeatism in his voice, for when he looked at the world, he found it so remote from his ideals. Zohra, sensing some disillusionment, said with a friendly laugh: 'The wheels may be pulling a golden chariot for you!'

'No gold or diamonds for me,' he laughed back. Happening to glance at his watch, he gave a start. 'It is past eleven! They will be anxious at home. Heaven knows what they might think.' He stopped abruptly.

'I hope your brother has been too absorbed in his work to think of the hour. He has a habit of worrying and fears all sorts of mishaps,' she said softly.

Hamid smiled, thinking how differently people acted when they were in love.

On reaching home, Zohra first went into the children's room; the sight of her children asleep, as always, sent a thrill through her. She kissed each in turn with a tender passion. Entering the bedroom, she found her husband had not yet retired. She quietly walked into his study. At the sound of her light footsteps, Bashir instantly rose, as she came towards him.

'I have been worried to death. Hamid was driving and he can be very absent-minded,' he said.

Without waiting for a reply he drew her to him fiercely, clasping her in his arms as if afraid of losing her. She made no effort to free herself. It was good to know that he still loved and desired her so passionately, even after five years of married life.

16

The house had been renovated before Bashir's marriage. It was now designed to accommodate both the sons with their families, in separate wings. The additions and alterations were in the western style, with modern conveniences.

Hamid was given the left wing, which for a bachelor was much too spacious. But it soon began to be filled with friends and visitors.

Hamid started his bookshop, a little way from the main city. He had found energetic well-wishers to help him in this venture. He also had a knack of making people feel at home, and his rooms soon became a haven for his friends as well as people in distress. But this lavish hospitality gave his more formal and class-conscious mother a severe headache. It also kept her busy, for she never knew how many guests there would be to cater for. She sometimes remonstrated with Hamid, but to no effect.

Zohra found a strange resemblance between Hamid and her own father. But her father's protégés belonged to the old school of thought, with leisurely ways, and were moreover grateful for his patronage. Hamid's crowd mainly belonged to the so-called Progressive Group. They were young, vital, tense, where ideas and ideologies were concerned, but easy-going and carefree where hospitality was involved, taking everything for granted. Two girls were also drawn into this set. Khorshed was tall and angular, and spoke rapidly with a guttural voice. She was an aggressive feminist, haranguing everybody about women's emancipation and rights. The other was Zohra's old school friend, Nalini. She had become increasingly conscious of the various political and social problems facing them, and though of a quiet disposition, found this crowd stimulating.

Zohra usually joined them whenever Nalini was present for, with her old-fashioned upbringing, she could not enter into the spirit of what seemed to her the ultra-modern ways of the others. Khorshed, with her dictatorial manner, was often brusque but her mind had a clarity and lucidity, which filled Zohra with awe. She had no patience with confused irrational discussions. 'Oh, do stop,' she would say. 'My mind is exhausted trying to follow the path of your convoluted brains.'

On one occasion Khorshed, stretched out on the divan, extended her legs so that they touched Hamid's, and said with an affected drawl: '*Arrey*, Hamid, get out of the way; I want to make myself comfortable!' It was the tone she always adopted towards Hamid, when not speaking of business matters. But while the others looked on amusedly, Hamid thrust her legs aside and gave her a smile that was a mixture of disapproval and appreciation. Khorshed, far from taking offence, only turned to him with an undisguised look of admiration, not unnoticed by Zohra. Then she shouted:

'*Arrey*, bearer! ... Hamid, now where is your bearer—dead? I want some tea.'

Zohra quietly rose to find the bearer, while Khorshed continued: 'This silly bookshop of yours, it's giving me a headache. You go on putting in money, and handing out salaries to the employees, and don't care a damn if it's ever going to be a paying concern. We'll all land in the bankruptcy court one day. You haven't the slightest idea about business.'

'Why should I, while you're here—our efficient business partner?' said Hamid half jestingly.

'What a fool you are,' said Khorshed tousling his head and pulling at a forelock affectionately, 'I don't know what you'd do without me.'

'Khorshed, you treat Hamid Bhai as if he were your *bap-ka-mal*,' teased Nalini, in her quiet way.

'Oh, Hamid needs to be indulged,' said Khorshed, coveting him now with her large expressive eyes.

Hamid lightly tapped her shoulder. He was the only one who did not take Khorshed's partiality to himself as signifying anything but friendliness.

'Besides, what right has Hamid to all this?' Her bare bony hand circled the air, but an undercurrent of seriousness ran through her words. 'Simply because it was his *bap-ka-mal*, his father's property, his *bap-ka-bap-ka-mal*, his father's father's property, and so on backwards for goodness knows how many generations? It's so old now that it stinks!' She sniffed, 'Such father's and grandfather's belongings are preposterous!'

'But until a better order is evolved, I'll hold on to them,' said Hamid quietly, 'whether you like it or not.'

'And anyway, does it follow that *you* get a proprietary right over them?' asked another, turning to Khorshed, with a sly dig at her fondness for Hamid.

'Oh, damn it all, why should that worry you? ... *Arrey*, bearer,' she called again, in her husky voice, 'what's become of my tea? *Sala*, is he dead?' she asked.

'Why do you call people *sala* when annoyed?' the same person interjected with an air of innocence. 'You brother will be the *sala* of the man you marry.' He glanced slyly at Hamid.

'What's it to you?' she asked. 'Amongst us we use brothers-in-law merely as targets for affectionate abuse. There may be some reason for it.' Her eyes almost unwittingly turned to Hamid, but his remained unconcerned.

Sometimes friends arrived without informing Hamid and stayed for days or weeks as it suited them. In spite of their superficial light-heartedness, these young people who surrounded Hamid were serious and earnest. At the moment, they formed only a small discussion group with big plans and limitless enthusiasm for religion, politics, sociology, economics, art, literature, and in fact almost any subject that touched their lives. One day they hoped to change the world.

Occasionally they arranged soirées of western music, for in Hyderabad, Indian music and dancing were left mainly to the professional dancing girls and had therefore acquired vulgar associations. Hamid had a good voice, which he had, to a certain extent, developed in Europe. But he was reluctant to sing to an audience, and it was only after strong pressure that he ever consented.

Bashir almost religiously abstained from entering Hamid's quarters. He also warned Zohra: 'Look, do not get mixed up with that peculiar crowd of Hamid's. They are out to destroy everything worthwhile.'

'I am not a child any longer,' she replied pleasantly, but firmly.

Zohra and Hamid seldom met alone, but she soon discovered that, although Hamid allowed his friends to collect around him and though he was interested and amiable, he often became ill-tempered and aloof. Their shrill voices and loud manners seemed to jar on his nerves; and even while they were in his house he would sometimes withdraw and leave them to themselves. But nobody seemed to mind his inconsistency. Sometimes for days together they found him totally absorbed in writing, when he would ignore everyone. But his friends seemed to understand, and loyally flocked back to him when the mood had passed.

There had been a week's spell of such quiet in Hamid's quarters when, one evening, Zohra, passing that way, saw him playing with her children in the jasmine arbour, and went over to him.

'Look, Ammi, look!' Shahedah, dancing around, proudly pointed to a tunnel made with Meccano pieces.

'And, Ammi, look at my train. Look, how it goes through—zoom!' Shameem gave it a push through the tunnel. 'Chacha and I made it.' Hamid had recently given them a Meccano set.

Zohra gave Hamid a smile of gratitude for his interest in the children. Hamid appeared as exuberant as Shameem and Shahedah themselves. Zohra joined in their game. After a while Hamid said:

'Zohra, let us go and sit out on the verandah. We can leave these two to play by themselves.'

There were loud protests from the children, but they were ultimately persuaded to continue playing on their own. Hamid promised to help them again the next day.

On the verandah, Zohra sat down and Hamid, pulling up a cane chair sat down near her.

'How is it you are alone these days?' asked Zohra

'Oh!' he said wryly, 'I was so irritable that my friends preferred to leave me alone. It's good to have a rest sometimes, even from one's companions.' He heaved a sigh of relief. Then loyally, 'I like them. Only, sometimes, I crave silence.'

'I was only looking for the children,' she said timidly, afraid that in this need for solitude, he might resent her presence too.

'Oh, I did not mean you,' he said, suddenly bursting into a welcoming laugh. 'On the contrary, you restore my energy. I have not seen you, that is, not *really* seen you for ages.' His voice and smile were genuinely warming.

Zohra, self-conscious, could only think of saying: 'I hope the children are not making a nuisance of themselves.'

'You know, I could easily get rid of them if they did,' he said simply. Then on another train of thought: 'You will have to watch Shahedah. With those eyes and that smile, she will be dangerous.' Glancing towards the arbour, he added: 'You will have to put her in *purdah* until she is respectably married.' Zohra's heart swelled with pride, but at the same time a faint blush suffused her face, as she remembered Hamid's remark about Shahedah's resemblance to herself. But Hamid had not spoken with the intention of flattering her.

'There is a long way yet for that,' she said as her eyes dwelt on her daughter, 'Who can foretell what will happen by the time she grows up?' Then she remarked, 'Your arbour will soon be flowering. It is full of tiny jasmine buds.'

'Yes, I was lucky to have this old creeper. It only needed a little tending. One of my roses too has just started flowering.' Rising from his chair, he said: 'Let me get you the first bloom.'

'No, no, please don't! Not at this hour.'

'Why not?' he asked. 'It would look beautiful in your hair. You are having guests tonight.'

'You know, all my life I have heard the elders say that one should never pluck flowers so late, and disturb the plant in its slumber. Perhaps it sounds foolish or superstitious,' she said, watching him, 'but I am reluctant to pluck them at this time of the evening.' She spoke haltingly, not knowing how her brother-in-law would take it.'

But he listened attentively, then remarked: 'Anyway, it is a pretty idea, and I should not intentionally flout it. We know so little of the life and emotions of anything beyond ourselves,' he said, looking towards the roses.

'I thought you once said we knew very little of our own lives too,' she countered, casting a mischievous glance at him.

'That is true.' His eyes earnestly searched hers, but they were now staring at the rug under her feet, as if she were studying the pattern.

It was getting late, and Zohra, calling the children, returned to her part of the house.

II

One day, Hamid, enjoying a respite from work and friends, suggested a visit to the Royal Qutb Shahi tombs. It was arranged he should take Zohra and the children there, early in the afternoon. His brother would join them later, direct from a university meeting. Bashir had consented to this plan only on Zohra's insistence. Having no desire for his brother's company, he had at first suggested that Zohra and Hamid should go by themselves with the children.

Safia, who had never been fond of such excursions, was going to the Zenana Club, to practise for the coming tennis tournament. Besides, the novelty having worn off, she often found Hamid's company irksome for she had discovered that mentally, they had drifted apart during Hamid's absence. Also she could not help feeling a little envy in her heart at finding Hamid and Zohra mutually interested in the sort of things that only bored her. So it was that Hamid went with Zohra and the children to the tombs. Leaving Shameem and Shahedah to play in the garden with their ayah, they wandered round, visiting the more important monuments, and reading the inscriptions.

As they halted on an open platform, Zohra, breathing in the air deeply, said: 'How strange that there should be no sense of tragedy or comedy here, but only of blessed peace. It is as if each monarch had found the repose he desired, although all must have struggled fiercely to live.'

'After every struggle, peace must descend and it is said that death makes everything ageless. But as the poet Ghalib wails: *Of what use are houris millions of years old?*' They both laughed. Then more seriously he remarked: 'Anyway, these tombs and the Golconda Fort give Hyderabad City its romantic background. They seem to provide the balance between the old and the new Hyderabad.' Hamid gazed at the old Golconda Fort, the capital of the Qutb Shahi Kingdom, all of whose rulers lay buried in these tombs.

The mausoleums, each erected in the centre of a raised platform and crowned with a dome, were scattered about in a fairly well-tended garden.

As they walked back towards the children, Shameem came running to Zohra, but Shahedah went straight to Hamid with tottering steps and outstretched arms. He lifted her up and perched her on his shoulder, unmindful of her shoes, which brushed his sherwani with mud.

As they strolled round the garden, they saw a *ber* tree and stopped. Shameem started picking up some fruit lying on the ground. But Zohra, noticing riper and fresher ones on a branch, leapt swiftly to draw it down. Hamid, watching her, thought she had the wild grace of a gazelle. She looked so young, her curls dancing on her forehead and her eyes sparkling like a child's. The bangles on her wrists, catching the glint of the sun's rays, shone more brightly and the texture of her mauve-blue sari, the colour of spring hyacinths, Hamid thought, appeared more delicate.

Little Shahedah, perched on her uncle's shoulder, plucked a *ber* and delightedly called, 'Ammi, Ammi!' As Zohra came closer, she handed it down, proud of her borrowed height.

With the inquisitiveness of a growing child, Shameem soon had his mind diverted by the slow rhythmic sound of water being drawn from a well by a pair of white bullocks. With his hands full of *ber* he ran towards it. The others followed at a more leisured pace.

Shameem stood and watched the placid-looking animals being driven by a *mali* up and down a hillock, drawing water from the well in a large leather bag. The bullocks ran down the slope with the bag full, and climbed up again with slow steps after the water had been emptied into a moat. From there it flowed into different channels, watering the trees and plants. Shahedah clapped her tiny hands and laughed with pleasure each time the water poured out in a sparkling cascade. She made a move to slip down from her uncle's shoulder and he lifted her down. The children watched fascinated. Seeing them thus absorbed, Zohra left them in charge of their ayah and, turning to Hamid, said: 'Shall we sit down somewhere and wait for your brother? He ought to be here soon.'

They sat down on the grass by a channel, away from the well, under a pomegranate tree, on the green branches of which hung fruit and deep-orange buds. The rind of some of the fruit had cracked open, displaying tantalizing ruby seeds.

As they watched the water flowing, Zohra said: 'This looks invitingly cool. We used to dip our feet when we were children.'

'Let us do that now!' exclaimed Hamid, reaching to take off his shoes.

He rolled up his tight trousers well above the ankles. They sat down side by side, dipped their feet in the crystal clear water and watched it flow past. Sometimes it came with a rush, then, gradually slowed down and flowed almost imperceptibly until another rush followed the drawing and emptying of the leather bag which held the water and controlled its release. Zohra's dainty feet were exposed beneath the clear water, the henna-tinted nails shining bright. Hamid discreetly marked their shape and hue. This was evidence of the leisurely life of a long line of ancestors, he thought, as he watched her gather up the pleats of her sari and hold them to one side to prevent the hem from getting wet.

'I can imagine my life slipping from under my feet,' said he, gazing into the water and speaking in a half-musing, half-jesting manner.

'What does it matter, if it could flow so refreshingly?' she asked, with a sidelong glance.

'Some people read the future in stars and palms; I shall read yours in this flowing stream,' he said in a playful mood.

'Read yours; you have your life still before you. My fate is already sealed into a set pattern.' She tried to laugh.

'You mean your life has begun to blossom, now that you have this pattern to follow,' he said.

'Begun or ended, what is the difference? I suppose it will go on weaving the same old design, which may grow dimmer and dimmer with age.' Her face was now turned to him. There was a rush of water, flowing with unusual force. It took them unawares, tilting Zohra's feet, and for a brief moment putting her slightly off her balance.

'I suppose this is what they would call "being swept off one's feet,"' she exclaimed thoughtlessly, but even as she said it, she experienced a strange sense of agitation.

'It merely shows the need to be cautious—to be on your guard,' he replied. His eyes seemed to carry a warning. 'A little stronger current, and you would have lost your balance completely.'

'I might like to feel the sensation. It must be exciting.' She was speaking without thinking, her eyes bright.

He burst into a laugh. 'You certainly have an adventurous streak in you. I did not know our Hyderabadi girls, under their outward calm, had this craving for precarious living.'

'Why, aren't they human?' she asked, still in high spirits. But suddenly her expression altered as she recollected her desperation in the face of her arranged marriage, which she had been powerless to avoid. In an effort to get over it, and not wishing Hamid to notice her confusion, she said, as she watched bubbles floating towards them. 'These bubbles are the hopes of life. Read them before they vanish.' She managed to make her voice sound light and happy.

'They speak differently for you and me,' he countered. 'For you, they signify the fulfilment of hopes and dreams. They dance a happy dance.'

'And for you?' she asked.

'Oh, for me! ... They play enticingly, they glow with a radiant light, and then suddenly vanish into the flowing stream.' He tried to sound unconcerned but she asked:

'Why are you so pessimistic?'

He now turned on her that intense look of his. Then bending a little, and resting his chin on his hand, said musingly:

'Happiness, in the present state of the world, seems as elusive as the sunbeams rippling on the surface of this stream. You know, if you try to catch them, they elude you totally.' Then, with a gesture of his shoulders that seemed to loosen his tense limbs, he heaved a heavy sigh. 'I should not talk such nonsense.'

'You have friends and admirers to share your visions. Out of this might be born something worthwhile,' said Zohra.

'But what have we achieved? It is often disheartening, wearisome; not friends of course, but everything else. In fact, life itself sometimes seems so futile.' He did not quite know why he was talking to her like this. It was just a passing idiosyncrasy.

'It should not, with Khorshed to keep up the tempo,' she said trying to restore him to a happier frame of mind. Nevertheless, she watched him keenly.

'Oh, Khorshed!' He gave his shy smile. 'Yes, she is a wonderful comrade, really like a sister.' He wanted to defend her against the suspicions that were prevalent.

'You mean like Safia Apa?' she asked, still teasing him. She had never considered his friendship with Khorshed as anything more than just that.

'Oh, Safia is such a harum-scarum!' he said carelessly, but immediately his face became grave and he lapsed into complete silence.

Zohra also remained pensively quiet. At last, when the silence threatened to become permanent, she said, reverting to the subject that had occupied them before Safia's name cropped up:

'Let us read your fortune in the stream again. Amma is anxious for you to get married; what does it say regarding that?' she asked with a questioning smile.

'It emphatically warns me against it.' He gave a mirthless laugh, then continued more seriously, his long fingers playing with the fallen leaves in the water: 'I feel it is not in me to make a woman happy. At heart, I am only a primitive, craving for the simple life. A small hut in the Himalayas, private and secluded. Which of your delicately bred Hyderabadi girls would care to live that way?'

'I am sure there are some,' she replied, carried away by the response in her own heart to her brother-in-law's unworldly idealism.

Hamid's lips twitched into a sceptical smile.

'I fear you are mistaken. In the eyes of most Hyderabadis, a man is worth little, save for his earning or inheriting capacity.' He was speaking slowly, 'If I were to marry now, I should be weighed on the same scales. In my case, it would be my inheritance and I should not then feel justified in changing my way of life.'

'Why are human values so wrong? I wish man had discovered some mode of subsistence other than money.' Zohra's voice was almost plaintive.

'It is all so hopeless,' he said, realizing that here at last was somebody who could understand his longings. 'People are expected to cultivate qualities that help them to acquire money and power rather than fulfil their ideals or purpose. The social order must change.' He abruptly turned his head and a lock of hair strayed onto his forehead, which he immediately pushed back.

'The accumulation of wealth for its own sake is a philistine concept,' Zohra mused. 'True contentment consists of so much else. Wealth could, at best, only add to one's comforts,' she said, watching his hand fishing out floating leaves from the stream. 'Happiness is more like the fragrance that emanates from within; as the poet Ghalib says:

> *It's true the eye doth not perceive the pierced heart;*
> *O, thou sceptic, canst thou see the blossom's fragrance with thine eyes?*

She looked pensive, as she quoted the lovely lines.

'Ghalib's poetry is transcendent,' he said, 'but one can use a modern analogy, for it is like the cosmetics used by women these days. They can enhance a woman's loveliness, but cannot create it. Sometimes they can even give the illusion of beauty; only it is not lasting.' Hamid burst into a spontaneous smile, idly expatiating on the subject, as his fingers wandered through his hair. Zohra knew this was a gesture he often repeated as if he did not know what else he could do with his hands.

'But you must marry,' she insisted. 'You are too young to waste your life in loneliness.'

As Hamid looked at her rather vaguely, she hurriedly quoted:

This world is ephemeral
And life perforce vanisheth,
Then whence this hesitation?
Do thou what thou intendeth.

'And so they lived happily ever after,' said Hamid in a cynical voice. 'To me, that sounds like an epitaph to one's individual liberty. I do not intend to shackle myself to marriage.'

'One might almost think you were talking from experience,' she commented.

'Perhaps I am,' he replied as if he were making a confession. 'Anyway, experience is better than ignorance; one at least learns what to avoid.' Then knitting his eyebrows, he said, 'Our morals are dictated by false values. Our lives have become so mechanized and rigid that happiness itself flies away in fright. I envy children their freedom from false conventions.'

Zohra and Hamid were both descended from feudal families, and had been brought up in ease and luxury. But Hamid had tasted privation, when his allowance in England had been stopped. It had only added to his disillusionment.

Zohra had never experienced hardships, but she thought she would not have minded them, for her brother-in-law's ideals pleased her.

Whilst they sat absorbed in their thoughts, another rush of water surprised them. Hamid's feet slipped, whilst Zohra's were again slightly lifted. It seemed to wash away their pessimism. Zohra's quiet laughter, floating over the bright ripples, was happy and Hamid readily joined in, saying:

'Zohra, your life stream is flowing entrancingly now.'

Suddenly Bashir arrived unnoticed. In their absorption they had not heard his footsteps. He had come by the back path and taken them unawares. He looked surprised at the sight, but asked pleasantly enough:

'What are you doing?' His question was addressed to both. The illusion of the moment was broken, for how could they explain their idle game to Bashir? Zohra looked up at her husband, as if playfully pleading guilty.

As Bashir turned to Hamid, something in his look roused Hamid to say: 'We were watching life slip from underneath our feet.' He was wiping his feet on the dry grass.

'I think it is time we stood up and acted, instead of sitting back to dream empty dreams,' said Bashir drily.

Zohra, who had also risen, saw Hamid's colour mount, his sensitive eyes flicker. She thought he was about to burst out; but with a quick glance at her, he only pressed his lips firmly together.

Zohra, feeling resentful at her husband's manner towards his brother, said pointedly, 'It was fun watching life slip that way.'

Bashir only gave her an indulgent smile and said:

'Anyway, Zohra, you have a cold; you should have been careful.' As she was also wiping her feet on the grass, he gave her his handkerchief, saying:

'You must dry them properly.' There was hardly any trace of the cold left, but Bashir was always overcautious with her.

Zohra wished her husband had not adopted such different attitudes towards Hamid and herself for the same offence.

'I am famished, let us go and get something to eat,' said Bashir, who had been watching Zohra dry her feet and put on her slippers. He was now close beside her, and putting his hand on her shoulder, he pushed her gently forward.

But Hamid, eager to get away even for a brief moment, said: 'You stay here. I'll go and get the picnic basket from the car.' Then, sensing the disapproval that this would provoke, he added: 'Or, if you like, I'll ask your driver to carry it.' He hastened away without waiting for a reply.

Zohra, watching Hamid's receding figure and the one now beside her, thought that even the clothes they wore were typical of their outlook. Hamid, on his return from abroad, had begun to wear Indian garments, whilst Bashir, even after so many years, stuck to the western suit, although collar and tie, shoes and socks, designed for a colder climate, were unsuitable in the heat of a Hyderabadi summer. Besides, she thought wistfully, Hamid at least wanted to identify himself with his surroundings, whilst Bashir, precisely efficient, still retained the sahib mentality. He still preferred to be driven by his driver rather than drive the car himself.

Hamid, now more than ever, realized that the three of them were incapable of enjoying anything together. Instead of Bashir being the connecting link between them, Zohra had become that link between Bashir and himself. He had wished to develop a brotherly attitude towards her, but perhaps these purely fraternal feelings only came through the long companionship of a shared childhood; and anyway, there was something in Zohra that made this feeling impossible for him.

17

Shahedah was nearly two years old when Zohra discovered she was expecting another child. As she was reclining on her divan in the verandah one day, sick and listless, Bashir, back from the university, ordered tea and came to sit by her side. Zohra looked preoccupied. Bashir, covering her hand with his own, asked:

'What is on your mind, Zohra? You look worried.'

'Nothing, except ...' she faltered.

'If only I could be sure that Safia Apa would make a good mother, I think I would have the courage,' she said, looking doubtful. 'For after all, if we do not give her our child, who else will?' She was also thinking of Yusuf as a father.

'If you wish to keep your child, no one has a right to take it away from you,' Bashir tried to comfort her.

'I know. But I myself feel torn with doubts, and you don't help me to decide,' she said wearily.

'How can I decide when it is you who want more children?'

'Two are not enough for a respectable family,' she said, looking at him smilingly.

'You are ambitious.' His fond eyes rested on her. 'I should have thought the family complete with a son and a daughter. Shahedah is delightful and everyone is captivated by her; she is like her mother. Shameem, they all say, is like me; at least, he has all my childhood stubbornness.' His voice had in it a ring of pride, for although he took little active interest in his children, he exulted in them. 'It is for you to decide, Zohra,' he urged, trying to sound dispassionate. 'Remember you once told me you wanted priority of rights over the children? You looked rebellious then.'

'Yes, but I don't feel so sure now. I was then young and inexperienced.' She laughed at her own immaturity. 'Living in this house has altered everything. Amma Jan loves them. Bhai Hamid adores them, especially Shahedah. How can I feel they belong exclusively to me? Even the children would revolt if I made such a claim.' She looked happy as she spoke of the love lavished on her children. Then she went on thoughtfully: 'But that only makes me feel a greater obligation to the family ... Oh, I feel so confused ... I wish someone would decide for me.' But even as she said this she knew that if anyone were really to try to settle it for her, she would protest.

A maid brought the tea tray, and Zohra poured out tea for Bashir. A glass of iced sherbet made of fresh *falsa* berries from the garden had been sent up for her, as she never could get accustomed to tea.

One day, Safia strolled in from her mother's side of the house and sat down beside Zohra.

Zohra, who was reclining on the divan, was feeling rather depressed.

'I wonder if it's really worthwhile having children. The process is hopelessly tedious,' she said, flicking, without reading, the pages of the book lying before her.

'Yet how soon you forget it once it's over!' exclaimed Safia, and at once became tense. Then she added: 'It's truly said, "Blessed is she who has the gift of motherhood." For what is a barren woman like, but a desert where nothing flourishes? Everyone looks askance at her, making it still more difficult for her to bear her lot.' Her voice trembled. 'Bhabi Jan, you should be grateful for the children when they come.'

Zohra was startled by the intensity in Safia's voice. She had not credited her with such depth of feeling.

'If you only knew how I have longed for a child all these years!' Safia drew in a deep breath. 'I remember when I first learnt that you were expecting a child, I felt sorry for you. It was so soon after your marriage, and to me you looked like a caged bird. But once the infant was born, there was such a joy of fulfilment in your face that I almost felt jealous. It's true what they say, a childless home is like a home without light.' She suddenly broke off, then exclaimed: '*Ai-hai*, I am talking foolishly!' She tried to laugh.

Zohra found her own heart beating in sympathy with her sister-in-law's predicament, but she continued to fidget with the book, opening and closing it aimlessly.

Safia continued: 'People think I am frivolous. But what is there for me to do? These last two years I have longed for a child more passionately than I can ever express. My husband too is restless. And he is handsome and charming!' A world of feeling was conveyed in that one sentence. 'A child would be an anchor. I can see why my stepmother induced my father to marry again, and also why my father was tempted to do so. I comprehend so much now that I did not understand before.' Her voice rose to a high pitch, then suddenly dropped. 'But second wives are out of the question these days. Besides, that experiment was not very successful for either of the women, nor indeed for my father.' She gave a dry mirthless laugh that reminded Zohra of Hamid. Ordinarily there was little resemblance between the two. She moved back and leaned against the bolster.

Safia had seldom talked to anyone so freely of her disappointment; now that the floodgates were open, she felt she could speak on for ever. Zohra was moved.

'I would go through hell,' Safia burst out passionately, pulling herself up straight again, 'I would risk life itself, if I could have a child! I should then at least die a fulfilled woman.' Her voice faded away, then quietly started again: 'But that is impossible now; and of what value is my life? Even God wouldn't know what to do with such a useless woman.' From her throat issued harsh, cynical laughter. 'If I could have adopted ...' Noticing the look in Zohra's eyes, even she felt abashed in asking her to make such a sacrifice.

A wave of deep compassion surged through Zohra. It was mingled with a current of thankfulness at her own good fortune, and out of this was born generosity. Aware of her sister-in-law's hesitation in asking what she knew was trembling on her lips, she volunteered:

'Safia Apa, do you want my ... our child?'

'More than anything else in the world.' Safia snatched at the words as eagerly as a lion its prey.

'You shall have it then.' Zohra's voice was fatefully calm.

Safia fell back against the cushions and in a voice choked with emotion said: 'Bhabi Jan, Allah will reward you for your generous heart.' She could speak no further for the tears in her eyes.

On hearing of the arrangement, Masuma Begum's hands went up in prayer. 'Allah be praised a hundred thousand times!' she said fervently. 'No woman was blessed with a more dutiful daughter-in-law.'

II

On a night when the moon was just a crescent and the stars shone brightly, Nawab Shaukat Jung Bahadur passed away in his sleep. The death was so peaceful that it almost bore the appearance of continuity.

For a time it left a lingering sadness in the house, not unmixed with the relief that he had at last attained what he had so long desired.

The estate had to be kept intact for the present among the mother and sons, and the joint household was automatically retained.

Safia's legal share was given to her—half that of each of her brothers, according to the Law of the Prophet.

Safia and her husband Yusuf were both extravagant and reckless. They had been living beyond their means and had contracted various debts; thus they were badly in need of the extra income.

Zohra's child, a son, whom they named Iqbal, meaning 'good fortune' was born at her parents' home where Zohra still preferred to go for her confinement. Safia, having made all the necessary arrangements for bringing up the infant, took him away on the sixth day. This early separation from the mother was considered essential by everyone, for the sake of the mother herself as well as for Safia. Even so, Zohra wept many tears, her arms aching to hold the child she had carried within her and who had lain near her for such a brief period.

When she returned to her husband's home she still had a pallor unusual after childbirth. Everyone showed her extra kindness. Her face reflected a kind of spiritual resignation, as if through willing sacrifice her soul had attained a new more profound depth.

Safia in those first few months could hardly bear to be separated from the infant. Almost bereft of all reason, she doted over him more than even Zohra had done over her first-born. Yusuf also looked more settled. Seeing them, Zohra was glad of her gift, hoping in her heart to have another, as soon as she felt strong again.

Bashir, meanwhile, was ambitiously planning a physics laboratory, the biggest and best equipped in India, where research students from all over the country could come together. He worked relentlessly even during term holidays, but sometimes became disheartened at the lack of response from co-workers and even superiors.

'This is Hyderabad with its leisurely ways. Everything must move at a tempered pace,' he thought with exasperation.

Zohra, not being fond of a club life, was often lonely. Hamid these days was either working frenetically or falling into periods of quiet brooding, when even his friends kept away. They called it Hamid's chrysalis period. At first, on such occasions he had welcomed Zohra's company; but gradually she found he was becoming irritable with her as well. Perplexed and hurt, once when talking to him of friendship, she remarked:

'Your friendship is like the sea waves—advancing in full confidence, then receding as suddenly.' She tried to hide an inner disappointment. But Hamid only looked at her with a strange withdrawn expression.

On another occasion, when she met him in the passage, he invited her to his verandah, and as they sat there he began to talk to her in the old, friendly intimate way, discoursing on books and authors. He also told her of his scheme for the publishing house that he intended to start soon.

Before they parted, Zohra remarked; 'You are strange. You either collect the whole world around you, or you become a total recluse. What else have you been doing?' She tried to sound casual.

He hardly looked at her, but fumbling for his pipe, said in a half-jesting tone: 'Trying to realize my obligations.'

'I thought you, at least, were free from them!'

'Is any human being really free?' he asked. Then with a shrug he went on: 'Except those, perhaps fakirs and sadhus who have sought deliberate liberation from life. For the rest, freedom just remains a beautiful dream.'

'If I were good enough, I would sketch you, lying with your eyes wide open to the heavens, and call it "The Dreamer". Then another, looking down with the same eyes towards the earth, and call it "The Dreamer Awakened".'

'Zohra, don't try to stir sleeping devils. They are best left dreaming.' Looking at her, he smiled his slow reticent smile that Zohra found so fascinating, and she remained silent.

Abdul, an old servant of the household, who now looked after Hamid's needs, appeared painfully dragging his stiff-jointed limbs and announced: 'Hamid Sahib, there is a man to see you.'

'Who is he?'

'*Arrey*, Sahib, one of those who come here more regularly than the new moon, with always a fresh tale of woe for Sahib's soft heart,' he answered, smoothing his straggling hennah-dyed beard. 'Somebody is dying, somebody is marrying, somebody is having a baby, and everyone needs Sahib's help.'

'You sound as if you had a grievance against people doing all this,' Hamid cut him short with an impatient gesture, embarrassed at having Zohra as audience to this recital. 'Get some sherbet for Dulhan Begum, while I get rid of him,' he said, affecting a carelessness he was far from feeling.

But as soon as Hamid had gone, Abdul came up to Zohra to talk to her. His long years of service with the family gave him the privilege of expressing his opinion about her brother-in-law whom he had looked after since his birth.

'Dulhan Begum,' he confided, 'with Allah as my witness, I tell you our Hamid Sahib was born a saint—a saint,' he stressed with his hand raised towards heaven. 'He opens out his heart and his purse to the poor, as no other man on earth does—no other man on earth—I tell you. You were not here to see, but in childhood he loved to dress himself in beautiful clothes. Even so, he could never bear to see anybody in need, without wanting to run to their aid.' The old man, with a gesture of his shaky hand, continued: 'To give has become a passion with him now. Dulhan Begum, if I try to prevent him, he says he gives because he no longer needs these grand clothes ... But his wardrobe is getting emptier and emptier, and so is his purse. He now goes about wearing those rough homespun shirts.' Then, hobbling a step closer and shaking his head in disapproval, he went on in a more conspiratorial manner:

'Dulhan Begum, you tell me, are such clothes to be worn by people of his status?'

Zohra could not help smiling, but she listened intently, reflecting on Abdul's words.

Taking encouragement from her reaction, the old servant continued: 'May Allah give our Hamid Sahib a long, long life, but if he feels that somebody needs anything badly, he even borrows to give.' Abdul, stroking his beard again exclaimed: 'May Allah preserve it, he has a large enough ancestral income, but how can things last? What with his bookshop and his friends and the eternal beggars. Even the treasures of Harun-al-Rashid would not suffice for such generosity ... Allah knows, Dulhan Begum, I am certain that he would give away his last garment and his last rupee if need be,' said the bearer with a final flourish of his gnarled hand.

As Zohra listened to the old man, her heart filled with a strange emotion, although she knew such indiscriminate generosity might often be misplaced.

Then, moving a step closer, Abdul spoke even more intimately, making Zohra uncomfortable as he bent over her, and she could not help drawing back a little. Holding his large turban in place with both his hands, he continued: 'Dulhan Begum, I think he will listen to you. Why do you not speak to him? He is not yet married, but what will he do when, Allah willing, he has a wife and children? Those so-called friends are all leeches—leeches, I tell you.' He nodded his shaky head emphatically, while his hand gesticulated.

'No, Abdul, I think you are mistaken,' said Zohra, with a grave inclination of her head.

Hamid's returning footsteps were heard, and the old man with a start said hurriedly: 'I must go and get sherbet for you, Dulhan Begum.' And he made a gallant effort to hurry away on his bare feet, for with his rheumatic joints he was slow and clumsy. But even in this state of infirmity, he would not give up dyeing his beard with henna, and altogether still wore a rakish look.

18

After the month of Ramazan, the Festival of Eid was approaching. Both Bashir and Yusuf would have a few days' holiday. Yusuf, who had a passion for fast cars, wished to try his new sports model out on a long-distance run. Having bought the car out of Safia's inheritance, he wanted to give her family a ride in it. Safia at first refused to go on account of the infant, but she could not resist Yusuf's importunity for long, and ultimately gave in. They agreed on a visit to the Ajanta and Ellora Caves.

But two days before the trip, Bashir said to Zohra: 'I am afraid it is impossible for me to get away. There is a lot of extra work I must get through during the holidays. Our people have no sense of responsibility. They put off things as if time will stand still at their pleasure. But somebody must do the work.' Bashir was quite apparently exasperated.

'Then I, too, shall stay. We can go together some other time.' She could scarcely hide her disappointment, however.

'You should not miss this opportunity, Zohra. You have always been keen on seeing the caves, and Hamid will be a good guide, that is if he does not get lost.' There was a hint of condescension in his voice, which Zohra ignored.

'I do not like leaving you alone during the holidays,' she said, clinging to her sense of duty.

'I shall feel easier if you go. I shall then be free to work.' He gave her an affectionate look and, scrutinizing her face, said: 'You need a change badly, too. You are so pale these days.'

Anxious to see the Ajanta frescoes, and absolved from scruples about her wifely duty, Zohra gladly agreed to accompany the party.

It was decided they would stop a night in Nanded. Bashir arranged to send telegrams and have the government guest house made ready to receive the party. The cooks, attached to the guest house, were instructed to prepare the evening meal.

They left Hyderabad at dawn. Masuma Begum had ordered a picnic basket to be prepared for the journey. Yusuf drove very fast, taking risks and enjoying his own rashness. The car jolted over roads that had suffered damage during the last monsoons. They all shouted protests, which went completely unheeded by Yusuf.

At last, when they halted for lunch, Safia threw up her hands, exclaiming, 'We shall all offer a prayer of thanksgiving for our deliverance. These high-speed cars have become the curse of our age. Look,' she said, addressing her husband, 'if you don't reform, we shall all disappear into nothingness.'

'That should please you. It would be a short cut to heaven,' retorted her husband with careless good humour.

'*Arrey*, and what about my child?' gushed Safia; but she sounded genuinely concerned.

'I would rather take the long way,' said Zohra, 'I am not prepared to meet my Maker yet!' Zohra's tone was always somewhat restrained when she spoke to Yusuf. 'You must have a guilty conscience!' Yusuf looked at her and gave a shrug.

Zohra, biting her lips, turned away, beginning to wonder if it was not foolish of her to make this trip in Yusuf's company. Hamid, taking offence for Zohra, said:

'In spite of what the scriptures say, I feel everyone has a right to take his own life, but certainly no one has a right to plunge other people's lives into danger.'

'Why, Bhai Hamid, *you* say this! I thought only ladies could be so panicky!' retorted Yusuf, with unnecessarily high-spirited laughter.

'Yes, even I,' replied Hamid. 'Although you may not consider my life to have equal value to yours, I certainly wouldn't like it to end in this senseless manner. But, anyway, you should consider the zenana.' He turned to Safia and Zohra.

'It's you who make them nervous.' Yusuf's look was directed more towards Zohra. 'There's no more risk in this than in scores of other things we attempt daily.'

'If there is an accident, the verdict would be suicide in your case, and murder in ours.' Hamid's voice was that of quiet acceptance of an unpalatable fact.

Yusuf only laughed unconcernedly: 'Oh, you and your pessimistic outlook! You are only fit to retire into a monk's cave. Ajanta is a good place for you!'

Hamid gave a sigh of resignation. It was useless trying to argue with his incorrigible brother-in-law, whose zeal for speed was insatiable.

That night they stayed at the guest house in Nanded, and left very early the next morning for a day's drive to Aurangabad.

II

On the edge of a crescent-shaped range of hills lay the caves, hewn deep into the rock, their walls decorated by a succession of devout Buddhist painter monks who had laboured over centuries at their self-appointed task. Time had tarnished much of the original beauty of the frescoes, yet some retained an astonishing freshness.

After a hurried picnic-lunch at midday beneath the trees, the party rested in the colonnaded entrance to a cave shaded from the sun. There, the guide drew the ladies' attention to a lotus design on the ceiling. 'The lotus is a symbol of purity and goodness,' he explained, 'and the cycle of buds and flowers represent the eternal cycle of birth, life, and death.' He went on: 'The Nizam's elder daughter had this design woven into a sari border. Ever since this fashion was set by the princess, there has been an increasing demand for saris with Ajanta designs.'

Safia's face at once lit up with interest, and the guide continued: 'You would know, of course, that nowadays it has become the trend for Hyderabadi ladies who come here to order such saris.'

Safia was enthusiastic, for although indigenous industries did not interest her, the princess's patronage gave this particular one the distinction she always sought.

'I'll buy some material like this to make a sherwani for Iqbal,' she said turning to Yusuf, completely unmindful of the unsuitability of the design for such a young boy. But Yusuf was able to suppress her foolish fancy.

Zohra was entranced as she wandered through the caves looking at the paintings of the Jataka Tales, stories of the previous incarnations of the Buddha. Their rich genius left her wondering what their pristine grandeur must have been like. The inscription on one of the caves exemplified her own feelings. The guide told her what it meant: *May the entire world, having got rid of its sins, enter that peaceful state which brings freedom from sorrow and pain.*

After supper, as they all sat on the guest house verandah, Hamid tried to explain the technique and skill which went into the making of the frescoes. He told them that the wall had to be prepared with a plaster of earth, sand, husk, and grass, followed by a layer of mud and crushed rock bound together with glue. Finally a coat of limewash completed the process, and once it was dry the painting could commence.

Safia and Yusuf soon became restive and retired.

But as a parting shot, Yusuf, turning to Zohra, said: 'I'll now leave you to your undiluted dose of art.'

Zohra forced herself to say goodnight.

After they had gone, Hamid said: 'A harmless person really, this Bhai Yusuf.' Zohra made no reply for she was not certain she could agree. After some reflection she said: 'What perplexes me most is that these frescoes should have been accomplished by monks. How can asceticism and art go together?' she asked, leaning back in her cane chair.

'Art inspires an artist in very much the same way as an ascetic is imbued with Divine Love. They are both derived from the same aesthetic stream,' said Hamid as he leaned forward in his chair. 'A true artist longs to attain oneness with nature, with humanity, through his art, whilst an ascetic seeks unity with the Divine through meditation. They both crave for the same Truth, the meaning of life. Therefore, the ascetic is often only a super-artist.'

'Perhaps you are right.' Zohra was trying to weigh his words. Then still not quite convinced, she asked: 'But have not the monks shown remarkable skill in painting women?'

'Yes, but they are portrayed here mainly as the most inspiring of human influences, through which a man may seek divine affinity, and not as temptress,' said Hamid very slowly, as if something was oppressing his mind.

'But how did the monks acquire such knowledge of women?' she asked with a half-challenging smile. 'These paintings might have been executed by some woman-worshipper rather than by a woman-shunner.'

'The monks were not born ascetics.' He laughed a little nervously. 'Probably, most of them were merely disillusioned lovers. Artists have acquired a reputation for inconsistency. But what appears as fickleness at first is perhaps something that touches a much deeper chord.' He now spoke earnestly as he lit his pipe. 'Endowed with sensitive souls, they are extremely susceptible to a woman's beauty and influence, setting her up as the highest of God's creation. Thus an artist is, forever, in search of perfection, an ideal union. In this, he is usually disillusioned. Failing to find perfect unity with a woman, he often renounces the world and seeks affinity with the Divine, trying to discover some hidden spark with which to ignite his being. Yet, what he missed in life, perhaps his soul still seeks in his meditations. Thus these artist-monks have given woman that inspiring grace and loveliness which they sought in vain in real life.' Hamid expressed his thoughts as if the subject were very near to his heart. He continued: 'I think that the love between man and woman that seeks its culmination not only in physical, but also in intellectual and spiritual unity, is the perfect symbol of an ascetic's passion to merge his self in the Divine.'

'Does Sufi poetry not express similar ideas?' she asked.

'Yes, and therefore the Persian love songs are full of burning passion, or rather they have a fervour, an ecstasy, far transcending human passion,' replied Hamid. His nervous fingers passed through his hair, pulling at it. As he met the soft intent expression in Zohra's eyes, he looked away and said abruptly: 'I did not realize how late it was. You must be tired.'

Before she could reply, he rose and walked away. Zohra was offended by this scant courtesy, as she herself had lost all count of time, talking to him. Hamid walked out into the garden, seeking the coolness of the fresh air and the solace of the stars, which shone bright in the night sky.

Next morning, they visited the caves again. Safia's and Yusuf's interest soon waned, and seeing the other two still absorbed in a discussion on spirituality, they decided to go around the caves at a quicker pace by themselves, But before leaving, Yusuf said to Zohra in a tone not quite audible to the others, 'I leave you to your enjoyment, Bhabi Jan,' and walked away.

Zohra gulped at her brother-in-law's inference.

There were some masterpieces, which stood out from the rest, and to these Hamid and Zohra returned several times. But Hamid appeared preoccupied and seldom made comments. One painting in particular seemed to draw them both to it. It showed the Buddha, having received enlightenment after seven years of wandering and hardships, returning to Kapilavastu, the city of his forefathers. His wife, Princess Yasodhara, with their son Rahula, is wearing her loveliest clothes and richest ornaments to welcome him back. The majestic form of the Buddha, his noble head slightly bent, looks down with the tenderness of divine love and humility at his wife and son, drawing them into the circle of his spiritual radiance and imparting to them the strength that he himself has gained. The sacred lotus flower lies at his feet and an angel floats in the air holding a flower-decorated umbrella over his head. At the sight of the glory of the face and bearing of the man who had once been her earthly lover, a divine spark enters the soul of Yasodhara and kindles her being. She bends low in obeisance and offers their only son, the heir to the kingdom, as a disciple to this man who is a stranger to his son, and yet is at once his earthly as well as spiritual father. In this painting was the final triumph of the Buddha as well as the monk-artist who painted it. The sight of the lovely Yasodhara, the wife whom he had once desired with all the passionate ardour of a young lover, and whom he had found hardest to renounce, now filled him with only a sense of deep spiritual love and exaltation.

Hamid and Zohra stood gazing at it, as the full glory of that great renunciation dawned upon them. This painting seemed to possess the power to impart to the beholder its depth of emotion.

At length, when Hamid turned to Zohra, he was struck by the look on her face, for in her eyes, those of Yasodhara had found reflection. The orange sari, falling in its simple lines along the rhythmic grace of her form, made her harmonize with her surroundings. The intense expression vanished as her eyes met his. She gave him a faint smile of recognition, as if a figure from one of the frescoes had been imbued with life.

'We ought to be going, it is late,' said Hamid, glancing at his watch.

'I wonder if Safia Apa and Bhai Yusuf are waiting for us,' she said with a sudden twinge of conscience. They cast a last lingering look at the panel and went out. The other two had already left.

As they looked down at the valley, their eyes were arrested by the evening glow reflected on the fields studded with wild yellow flowers whose heads were turned towards the sun. It was as if the plants, long drenched in rain, had put forth their blossoms and transformed them into sun-worshippers.

'Shall we rest here a moment, before going down?' asked Zohra, breathing in the fragrant air, and always anxious to linger in his company.

'Yes. It is cool and refreshing here.' He was feeling more at peace now.

Zohra sat on one of the stone steps, whilst Hamid, leaning against the trunk of a neem tree a few yards away, lit his pipe. After a long silence, he tapped the pipe against the branch and said: 'That painting is a masterpiece. It is the culmination of all that the poet Edwin Arnold expresses:

> *Oh summoning stars, I come! Oh mournful Earth!*
> *For thee and thine I lay aside my youth,*
> *My throne, my joys, my golden days, my nights,*
> *My happy palace and thine arms, sweet Queen,*
> *Harder to put aside than all the rest.*
> *Yet thee, too, I shall save, saving this Earth;*
> *............................. Now will I depart,*
> *Never to come again, till what I seek*
> *Be found—if fervent search and strife avail.*

Hamid's voice was ideally suited to bring out the pathos of this passage.

'The Buddha's renunciation is unique,' said Zohra, obviously moved. 'Nothing could be more inspiring. One feels it all the more these days, when the so-called great men of the world are so greedy for power.'

'The philosophy of life today is one of utter selfishness,' Hamid commented sadly.

'And all that we need is peace—external as well as internal peace.' She took a deep breath.

'No one can achieve that internal peace without a struggle; even the Buddha had to struggle before he could choose his path.' Hamid's look was one of intense concentration.

Zohra turned towards him as if wishing to hear more, and he went on:

> *This is the night!—choose thou the way of greatness or the way of good;*
> *To reign a king of kings, or wander lone, crownless and homeless, that the*
> *   world be helped*
> *I lay aside these realms*
> *Which wait the gleaming of my naked sword*
> *My chariot shall not roll with bloody wheels*
> *From victory to victory, till Earth*

> Wears the red record of my name. I choose to tread its paths with patient,
>   stainless feet.
> ....And all my soul is full of pity for the sickness of this world;
> Which I will heal, if healing may be found by uttermost renouncing and
>   strong strife.

There was a hush in his voice, vibrant with the deep emotion in the words. He rested the back of his head wearily against the tree. They both remained silent in this mood reminiscent of people who are honouring the dead. It was, at last, broken by Zohra, asking:

'Do you think it possible that Mahatma Gandhi is a reincarnation of the Buddha sent to cure India's ills once more?'

'How can one say whether he is an avatar of the Buddha or not; but in my opinion he is undoubtedly a saint,' said Hamid, turning to Zohra. 'Look at the way his ahimsa has gripped the imagination of our people. Look at the way the simple folk fall down at his feet to worship. There is something in him that commands such reverence.'

A quiet fell over them again, and again it was Zohra who broke it:

'It is amazing how Edwin Arnold, Englishman though he is, has caught the Buddha's spirit with such understanding.'

'It would have been good for both the peoples had there been more poets like him, instead of the Kipling variety, so weighed down with the White Man's burden.'

And Hamid moving away from the tree trunk, quietly sat down on the thick root jutting out from the ground, about a yard away from Zohra, and was putting away his pipe, when turning to him, Zohra asked:

'Why is it that there is not a single artist in our country today who can create anything half as beautiful as Ajanta?'

'The last two hundred years have been the most sterile in our cultural history. The creative powers of a nation suffer disastrously under any foreign domination. We were far too busy imitating our masters. Western things were imposed upon us with a complete disregard for our old heritage. The Mughals, at least, made this their home. They married Rajput princesses. Their culture mingled with ours in a fresh stream. Look at Mughal art and architecture.' Hamid was worked up now, as his hand gesticulated.

Having been educated in England, he had an understanding of the

British and also a regard for them. But he felt a keen disappointment that an opportunity to create a new equal society had been lost. To cheer him up Zohra said:

'Anyway, we Muslims possess the proud heritage of three different cultures—Indian, Arabic, and Persian—flowing through our veins in one tumultuous stream. Hyderabad, at least, has tried to retain parts of all these and to add the new western, as the most vital stream in present-day life. We shall yet achieve that unity of East and West—that much-desired synthesis.'

She threw him a challenging smile, with a proud toss of her head, in the modelling of which all the traditions and conservatism of the different heritages had come together in varying degrees. Hamid, watching her, asked:

'But is not that threefold pride in our past enough for our humiliation today? It is we who have allowed our heritage to lapse into this state. We have gone a long way from the days of Mohenjodaro, haven't we? Possessing one of the oldest civilizations, dating back to over five thousand years, we have reached this glorious state of bondage. Some other kind of cooperation would certainly have proved beneficial to both peoples. Science, if rightly used, has vast possibilities.'

'What do you mean by "rightly"?' she asked.

'Well, you know, we must make machinery our slave, not our master. We must use it, not be abused by it. Sometimes one feels like throwing all the scientific inventions into the sea. They have caused such desperate human misery.'

Zohra, still mentally comparing the present with the past, asked after a while: 'Is it true that even the Taj Mahal was once considered so worthless that the British were willing to have it demolished and sold for the mere value of its marble?'

'As far as I know, it is perfectly true, and I think this represents the sum total of the mentality with which they approached everything Indian. But we must admit that there have been great Englishmen, scholars and thinkers, who have helped to make us once more conscious of our great heritage, even before Mahatma Gandhi woke the mass consciousness,' said Hamid, speaking in his slow voice.

'Rabindranath Tagore will give new impetus to the renaissance of the arts and Shantiniketan will lead us back to our creative heritage.' Zohra looked wistful as she thought back to the days before her marriage when she believed she might have been allowed to study art at Shantiniketan. She now confided her girlhood longing to Hamid.

Hamid did not speak, but his eyes met hers searchingly. Zohra suddenly looked away, and a constraint fell over them both. Hamid had decided long ago, that Zohra's love for her husband, though calm and unwavering, had left the depths of her nature untouched. He had also marked that, although there was often an intense spiritual look in her eyes, there was also something deeply passionate in them. He knew that a nature like hers could find fulfilment, either in an all-surrendering love, or in complete renunciation. He now felt that if he so wished, he had the power to rouse those depths, which would destroy them both, robbing them forever of all peace. Hamid was looking intently in the direction of the valley, but his eyes were for the moment blind to it. Zohra, too, was disturbed. She was thinking of her life and her girlhood dreams, the study of art, the desire for a different kind of marriage. At this last thought she unconsciously turned to Hamid. The slim immovable figure, seated there, with his long slender fingers tensely clasping one of his knees, struck a chord somewhere deep within her. But as he was looking away, she could only see the sensitive profile, with its classic mould of the head, and he seemed to her the very embodiment of romance. Suddenly another long-forgotten figure flashed through her memory. But, whilst Siraj had, with every turn, every word, every gesture, tried to make love to her, Hamid had never behaved in any way other than would be considered right and seemly by the strictest of codes.

The realization crept into Zohra's being even as the dawn silently floods the heavens, that Hamid had a fascination for her beyond what she had felt for any other man. Their relationship, together with her own married state and her strict upbringing, had never allowed her before to admit this even to herself. She was both thrilled and alarmed at the thought. But what about him? What kind of feelings did he entertain for her? If he tried to make advances towards her, how would she behave? Startled, she tried to pull herself together. Sometimes it was good to have mental inhibitions. Until now, she had felt so secure.

It was Hamid who, waking up from his reverie, said: 'It is late. We should have gone back long ago.'

III

That night at dinner, they were both self-conscious, trying to keep up a forced conversation with the others. Suddenly Yusuf exclaimed, '*Arrey*, Bhai Hamid, I'm sick of all this arty talk. I am now going to relax with a whisky and soda.' He cast a quick knowing look at Zohra, which did not go unnoticed by Safia, and grinned broadly as if he had made a most witty remark.

Hamid, though no teetotaller, refused to drink. Yusuf's tongue was loosened as his senses became animated. He was in his element this evening, recounting spicy tales of questionable taste to his audience. Safia listened with some apprehension, as Zohra looked increasingly ill at ease, for Yusuf seemed to take a marked delight in embarrassing his sister-in-law. She was seated on the sofa with Safia, and Yusuf had drawn his chair next to her, while Hamid sat near Safia, their chairs arranged in a semi-circle. The drinks stood on the table close to Yusuf.

When at last they all retired, Yusuf turned to his wife, muttering:

'What's the matter with your brother? Why can't he drown his disappointment in drink like a man?'

'What do you mean?' she flared up, cutting him short, 'Why should he drink? You drank far too much this evening and spoke too freely in the company of ladies. Hamid has no craving for whisky as you do!'

'No, he has only a craving for other men's wives!'

'You sound as if you were jealous!'

'Jealous! Jealous indeed! I don't know what your brothers see in her! She's just a common ...'

'Stop! You are intoxicated!' Safia's anger burst forth, half drowning the rest of Yusuf's sentence.

'Yes, so I see everything double. I can now see those two staging a play for Hyderabadi audiences. For her, it's just dramatics.' He made hazy gestures, then laughed animatedly.

'You have a vile mind,' said Safia, for once thoroughly roused against her husband and alarmed at the viciousness of his remarks, 'Bhabi Jan is not like you. I don't know why you are so prejudiced against her.'

'You're blind.' Yusuf blinked his slightly glazed eyes. 'Anyway, I enjoy such situations. They are exciting.'

Again, there was a guffaw of laughter. Then sidling up to her he leered, 'Shall I tell you a secret? She tried all that on me and failed. Don't you see how cut up she is with me now?'

Safia was aghast. 'You are a teller of monstrous tales,' she said, trying to control her voice, but her tone belied the doubt that was beginning to creep into her mind. Was this one of Yusuf's inventions? Had he concocted this story because it was he who was attracted to Zohra? She brushed away the thought as quickly as it formed and turned her eyes on him. 'What you say cannot be true,' she declared aggressively.

'What can't be true? When I say it, it is true!' retorted Yusuf offensively.

Safia's head was in a whirl. She felt hysteria mounting and she clapped her hands over her ears as if to shut out Yusuf's accusation. Zohra, who looked so modest, was it possible that she could go about vamping men? Yet it was easy for Safia to think of Zohra being attracted to Yusuf, for with her passion for her handsome and dashing husband, she easily imagined that no woman could help falling in love with him. Safia was confused. Confusion led to imaginings and then to certainty, so that by the time she dragged herself to bed, she had convinced herself that what Yusuf had recounted was the truth. Although aware of his weaknesses, her jealousy towards Zohra allowed her to believe him completely.

Safia could now see quite clearly why Hamid and Zohra were so often absorbed in conversation that, to her mind, would not interest any normal person. But the blame fell on Zohra; for Hamid she only felt a great pity.

In the morning, as soon as Yusuf got up, Safia said to him:

'I hate the very sight of her! I didn't know we had a serpent in my brother's home, in the guise of a poisonous vine; and that she should be the mother of our child! I shan't let him go near her!'

'Don't be stupid!' said Yusuf. 'It's below your dignity to give the affair undue importance. Only, it's better to be on your guard, and not to believe all that she says, as if she were an angel.'

'Oh, I can't, I can't!' exclaimed Safia vehemently. 'Allah knows, I believed in her as in no one else. But it would seem that while I looked trustingly into her eyes, she was trying to cut the ground from under my feet.' Safia was deeply distressed.

Yusuf was now really alarmed at the repercussions of his thoughtless words. He had to exert all his power over her before she would promise not to make a scene and try to behave normally.

But henceforth a tension was apparent in the relationship between the two women. This, coming so soon after her child had been taken away from her, hurt Zohra all the more. Safia impulsively started being over-affectionate to Hamid, as if trying to woo him back into the fold of his own family but he, not understanding her motive, was somewhat irritated by such obvious excess.

From Ajanta they made a trip to the Ellora Caves, where Buddhist, Jain, and Hindu sculptures were carved in stone with masterly skill and endless patience. Looking at the figures of Rama with his devoted wife Sita, Safia could not resist the temptation of saying:

'Why is such stress laid on constancy in women? They often are no more constant than men!' Although not looking at anyone in particular, she seemed to be attacking someone. Poor Safia had never been faced with temptation and could, therefore, feel supremely righteous. But Yusuf was moved to comment:

'Man is the privileged being. But constancy is an overrated virtue, anyway.' He laughed with theatrical heartiness, and turned a half-antagonistic look on Zohra, but she had already turned back to the sculptures. Safia, closely watching them both, took this as further proof of Zohra's guilty conscience.

19

After this holiday, Hamid plunged more deeply into work. He seldom joined the family even for dinner.

One evening he walked in late, seemingly in a hurry, to see his mother. The old lady's pent-up anger burst out:

'*Ai-hai*, Hamid, what has happened to you? One would think you were the chief *Vizier* of Hyderabad, the way you keep busy'.

Hamid laughed without amusement. 'Amma Jan, perhaps we ordinary people have to slave more over less important matters.'

'But what for? Even if you did nothing, you could live comfortably on your income. Instead, you run a mad bookshop and waste your energies. *Ai-hai*, and what a way to spend money!' She scrutinized his khaddar clothes. 'Do these handspun clothes become a Nawab's son?' She sounded both disapproving and worried. The servants had been carrying tales.

'There is surely no reason for you to make yourself so unhappy about me,' protested Hamid with sudden fierceness.

'Your temper also has become intolerable. Even the servants are upset. You know they are deeply attached to you.'

'Yes, I think I need a change,' he said, sitting down at the edge of his mother's divan. 'I shall go away for a short time.'

'Yes, go away, go away, as if that were the remedy for everything!' his mother burst out. '*Owi*, what is the sense of collecting that crowd around you? They are vampires. And that girl Khorshed, everybody but yourself can see that she has her eye on you!'

'Yes, Amma Jan, everyone seems to have an eye on me. I am so utterly desirable!' He made an impatient gesture, wanting to get away from this conversation.

'*Ai-hai*, Hamid, son, listen to me, you are all keyed up about this Gandhi's Satyagraha. What have we in Hyderabad got to do with it?'

'That is our difficulty. We imagine ourselves in a world apart. But, after all, we are part of India and what affects her will eventually affect us. We must realize this,' he declared emphatically.

But his mother refused to take him seriously. 'Son, I may be lacking in understanding, but why must you worry about everything?' She was looking at him earnestly. 'I think you need to marry some lovely girl like your Bhabi Zohra. Look, Allah be praised, how happy your brother is!'

'Amma Jan, am I to hear of nothing but marriage? And why should I marry someone like Zohra?' Hamid flared up.

'Why, Hamid, what is the matter with you? I thought you liked her!' His mother looked pained.

Hamid, composing himself, tried to calm down. He went over and affectionately placed his hands on his mother's shoulders.

'Please make a paan for me and tell me what you have been doing. I am sorry for having neglected you of late, but such frightful things are happening in British India.' He looked shocked and thoughtful.

As his mother prepared the paan he tried to talk to her in his old way, but in his heart he wondered how a shrewd woman like her never guessed.

But this set him thinking seriously about what had lately been passing through his own mind. He had realized for some time past that Khorshed was attracted to him and would probably consent were he to propose to her even at the risk of alienating her own people. But there had been quite a few Muslim-Parsi intermarriages. As for himself, the religious difference did not matter. And she would at least be an intelligent companion sharing his interests. Often he thought he would broach the subject but, face to face, in spite of all encouragement from her, he could never bring himself to do it. He sometimes thought, why not just have an affair, if she were agreeable. He could think of all kinds of reasons for two modern young people like themselves wishing to retain their individual freedom and being reluctant to settle down to marriage. But for him, he well knew, these reasons were not the real ones, and Khorshed was too fine a person to deceive in this way.

Hamid was consumed by these thoughts which, day or night, gave him no peace. They made him nervous and anxious, and his friends began to notice his extreme agitation at the slightest controversy.

Memories of the trip to Ajanta made Zohra more restless too. Safia's behaviour had also changed. She seldom brought Iqbal with her even when she called to see her mother. When Zohra made enquiries about him, she retorted with caustic remarks. One day, happening to meet her alone in the courtyard, Zohra could contain herself no longer:

'Safia Apa, we gave you our child because you wanted him so badly. He is yours now, but why do you want him to grow up a stranger to us?'

'I did not know you would keep on reminding me that the child was yours!' Safia flung back. 'You may have borne him, but I have taken all the pains to bring him up. You have other diversions!' Safia added, bitingly.

'What has happened to you, Safia Apa, why do you talk like this?' asked Zohra, holding back her tears.

'Ask your own heart!' Safia's self-righteousness revealed her own complex frustrations and jealousies.

Zohra could make no reply. But she was greatly moved to find that her mother-in-law had become her staunchest supporter. In fact she never let an opportunity pass without trying to impress on her daughter the selfishness of her behaviour.

One day when Safia was visiting her mother, she sternly remonstrated: 'Safia, your attitude to your Bhabi shames my grey head. I could never have believed, had I not seen with my own eyes, that such ingratitude could dwell in the heart of my own flesh and blood.'

'She has bewitched you as she has bewitched so many others,' Safia did not disguise her insinuation, 'and as she had bewitched me. But you will find her out some day, even as I have done!' she added heatedly.

'Allah forbid, Daughter! I shudder for you. You have taken her child, and this is how you reward her!' The mother's eyes were turned upwards as if she were asking forgiveness for her daughter.

'The child is in better hands,' said Safia with smug conceit.

'Even though you are my daughter, I cannot agree with that. *Ai-hai*, Safia, you behave in such an impulsive and unbalanced manner. And your bridegroom, even he does not check you. In fact, he seldom comes to see me these days, since I started remonstrating with you.' Then, looking at her with a warning in her eyes, she observed: 'Somebody has been poisoning your mind. Our Dulhan is a pearl.'

'You can keep your pearl. I want none of its lustre!' said Safia, and stalked away awkwardly in a huff.

The tension between mother and daughter grew. Safia's visits became more and more rare. But this only made Masuma Begum increasingly loyal and affectionate towards Zohra, who was looking lonely and forlorn.

Bashir usually returned late in the evenings. He was now principal of the Science College, and had a great deal of extra work to do and meetings to attend. He worked hard himself and tried to make others work. He was respected and held in awe rather than loved. Of what was happening at home he had little knowledge, except that a rift had developed between Zohra and Safia. He asked Zohra not to take notice of his sister's eccentricities.

Zohra sometimes heard a babble of voices from Hamid's quarters, but even when her friend Nalini pressed her to accompany her, she lacked the courage, and Hamid himself nowadays seldom asked her to come. Once, during an accidental encounter, she admitted to Hamid: 'I wish I could do something constructive and useful.'

'You should take up some work,' Hamid counselled. 'Why don't you teach in one of the smaller schools. They need honorary teachers.'

'I was thinking of something of the kind, but your brother does not like the idea,' she said, merely stating a fact.

'Then start a class at home. Get poor children to come to the house. Teach them to read, write, sew, do handwork. No one can object to their coming here. It is a very genteel way of working.' Hamid could not keep the sarcasm out of his voice. His brother's ideas of class dignity were becoming more and more intolerable to him.

Zohra earnestly took up the suggestion and soon started such a class.

But one day came news that her father, who had been in indifferent health, wished to see her. She went, and finding him weak and depressed, decided to stay on. After a fortnight, as he was better, she wished to return home. But her mother remonstrated:

'*Owi*, Zohra, you and the children act like a tonic on your father, and where is the hurry to go back? Cannot your bridegroom do without you?' The Begum Sahiba sounded a little piqued. She felt Zohra was drifting away from them.

'No, Ammi Jan, it is not him. But I have to get back.'

'*Ai-hai*, I know, those little children in your school. Are you adopting them all?' The Begum Sahiba disapproved of such total absorption. Zohra only smiled wanly.

One evening, when Bashir returned home, he found Zohra unexpectedly back. As they sat down to tea on their private verandah, he said as he took her hand in his:

'You have made me so dependent on you that the older I grow the more lost I feel without you.'

She gave him a grateful smile, saying: 'But usually you are so absorbed that you hardly have time to notice my existence!'

Bashir defended himself: 'Even when I am unable to be with you, Zohra, you know I feel different, knowing that you are here.'

Then, feeling happy at having her back, and a little guilty at his own neglect, he decided, with an impulse unusual in him, to devote the whole of that evening to his wife. For, in spite of his preoccupation with work, she was to him still the most desirable woman, and he loved her deeply. The children took a secondary place in his life.

'You have not visited the university for a long time,' said Bashir. 'The new buildings are now complete and already in use. Shall we drive out there? I will drive. We need not take the chauffeur.'

'Yes, indeed.' Her eyes lit up.

'You know the vice-chancellor is to retire soon, and people have started saying that I should succeed him.' He looked at her. She knew he wanted her applause.

'I shall be proud to see you achieve the recognition you deserve!' she responded with genuine pride.

But he knew she could not fully realize the importance to him of the accomplishment of his personal ambitions. These things did not affect her in the same way.

The university buildings, scattered over a large area, were handsome and imposing, constructed in the neo-Indo-Mughal style of architecture. Some of them were still under construction. Bashir showed her around with much enthusiasm, for the university was his life. The completion of the building of Osmania University was the fulfilment of a dream that the Nizam of Hyderabad had conceived nearly a quarter of a century before. All the courses were conducted in Urdu.

The dignity of the buildings and their beautiful surroundings pleased Zohra. There were students playing on the sports fields, but although they greeted Bashir respectfully from afar, not one came over to meet him.

It was getting late.

'We ought to be returning,' said Bashir, looking up at the sky. 'The clouds are black overhead. We might get caught in a shower.'

'It may pass off,' she said, her eyes scanning the sky. 'Rain is not usual at this time of the year.'

As the car slid along the quiet road at dusk, Bashir steering with one hand, slipped his other arm around Zohra, trying to draw her close to him. He had only got her back that evening, and liked to feel her near. But with a slight movement she disengaged herself and said:

'Had you not better concentrate on your driving?'

'Why are you afraid?' he laughed, drawing her still closer.

'Be careful! Somebody might see us!'

'Few people are around at this hour.'

'Suppose some of the students did pass this way and caught their future vice-chancellor acting in this undignified manner!' she teased, trying to extricate herself.

'Yes, it would indeed be a shock to them! They hardly realize that professors are also human. They take us for old fossils, and imagine that love is the monopoly of youth,' he said, releasing his hold, however.

He was thirty-eight, and Zohra twenty-six.

'There is a scheme to start a women's university on similar lines,' he observed. 'The present girls' college has become too congested, although they are mostly first- and second-year students. You know how marriages interfere with our girls' studies.'

Zohra knew only too well that in this society women were permitted to study until a suitable bridegroom could be found. She thought of her own interrupted education and was overcome by a deep regret.

As the conversation lagged, Bashir's arm again slipped around Zohra. He held her now more firmly, laughing away her fears. She nervously strained her eyes and ears for vehicles that might be passing that way. Such emotional displays in public places were, to her, the height of immodesty.

The car gave a sudden lurch. Bashir quickly braked and got out only to discover a puncture. He at once set about changing the tyre. Zohra got out, too, and offered to help.

'This is not a woman's job,' he said firmly, gently pushing her away.

The sky was ominous; the overcast heavens looked as if they were only holding their breath before letting loose a violent torrent. Suddenly, before Bashir had finished, it started to pour. Bashir hurriedly pushed Zohra back into the car without heeding her protests.

'You should also take shelter until the shower passes,' she remonstrated.

'No chance of its passing off so quickly. Besides, I am already half-drenched. This will soon be fixed, and then we can go home and change,' he had to shout in order to make his voice heard above the storm, and he carried on with his work.

By the time Bashir got back into the car he was thoroughly soaked. Zohra, looking around, saw nothing she could use to shield him from the chilly wind. Stretching out one end of her sari, she tried to cover his wet back.

'How is this flimsy silk going to protect me?' laughed Bashir, though pleased at her concern. 'It is not easy for people like me to fall ill.' He was always proud of his physique. Nevertheless, Zohra felt him shivering quite violently through his drenched clothes.

That night he had a high temperature, which developed into pneumonia. It was as if some vengeful God had heard Bashir and was punishing him for his self-assurance. Gradually, his condition worsened and grew critical. Zohra sacrificed all rest and remained constantly by his bedside. Whenever she left him, even for a moment, he called out for her in his delirious state. The nurse could not quieten him. His mother, his relations and the servants, all wracked with anxiety, were only too eager to relieve Zohra and sit by the patient's bedside, but none other could soothe Bashir. Even Safia, in the face of tragedy, seemed to have put aside her antagonism towards Zohra and wanted to help, but it was ineffective.

One day Hamid offered to relieve her. Knowing the lack of understanding between the brothers, Zohra was at first sceptical. But, to her amazement, Bashir, in his semi-consciousness, found Hamid soothing, and whilst his brother was with him, did not call for her.

Thus gradually Zohra and Hamid took turns by Bashir's sickbed along with the nurse. Hamid watched Zohra's anxious face, her large eyes looking larger, with dark shadows encircling them. Tears, which she fought to control, often brightened them. Hamid sat in the bedroom night and day now, feeling guilty about the resentment he had sometimes cherished against his brother's very existence. It was a sort of penance that he was inflicting upon himself.

Bashir's illness had lasted for a fortnight, and just when they believed the crisis was over, the doctor revealed to Hamid that although the disease was under control, the patient's heart was sinking. He gave him injections, but the response was negligible.

Zohra's father, who had become increasingly devout, kept almost ceaseless vigils, in spite of his ill health, praying fervently for Bashir's recovery. He had not hitherto interfered with the treatment, but now he said peremptorily:

'Mian Hamid, Daughter Zohra, harken to an old man, and send for a hakim at once. Allah knows, their medicines for the heart have a magical effect. They have been tried scores of times and proved successful, where all western remedies have failed.'

Zohra turned anxiously to Hamid as if asking for his opinion.

'By all means, let us send for the hakim,' said Hamid. 'Brother's heart is not responding to this treatment, and I do not know what else we can do.'

They asked the doctor, who could not refuse.

On the Nawab Sahib's recommendation, Hakim Shaikh Afzal, one of the most renowned of his profession was called in. Sitting down by the bedside, he quickly measured up the situation and concluded that a cure was indeed possible. From his bag he produced little phials of powders and mixed three of these into a slip of folded paper. Then saying a short prayer he gently prized open Bashir's lips and slipped the powder onto his tongue.

The doctor, who was present, was watching for the effects with keen interest. For, although he had had some experience of such cures he could never be certain that it was not merely a fortunate coincidence.

Meanwhile, the Nawab Sahib, carried away by his great love for his daughter, sacrificially walked round the sickbed three times, offering his own life for that of his son-in-law. Although he murmured it was no fair substitute, he still prayed to Allah to grant him this favour. The idea of Zohra being under the shadow of widowhood filled him with unspeakable anguish and horror.

The fates were in Bashir's favour for he responded to the new treatment, and within a couple of days the crisis had passed.

'The Hakim Sahib has worked a miracle,' were the words on everybody's lips. Hakim Shaikh Afzal himself looked very conscious of his achievement, for it was seldom that he had an opportunity of showing off his skills in front of western-educated people, who seemed to have lost faith in their own heritage.

'Hakim Sahib, may Allah bless you!' exclaimed the Nawab Sahib fervently. 'You have saved our family from darkness and ruin. You have saved my daughter's home.'

'What could I have achieved, Nawab Sahib? I was only a weak instrument in the hands of Allah,' replied the hakim with an elaborate show of mock humility.

The Nawab Sahib sensed that he was waiting for more plaudits and as he was in the mood to accord them, he observed solemnly:

'Hakim Sahib, Allah has specially gifted your hands with the power of healing, and although some can acquire such powers, there is none comparable to you!'

'Allah be praised for bestowing upon me such a gift.' The hakim raised his arms in prayerful gratitude.

20

When Bashir could move about a little, it was arranged he should go to a friend's country-house to convalesce.

Zohra was naturally to accompany her husband, but it was also considered advisable by the elders for Hamid to go with them, for the cottage at the foot of Mir Mahmood's Hill, although only ten miles from the city, was rather isolated.

When Safia heard of this, her attitude towards Zohra, which had softened somewhat during Bashir's illness, hardened again, although Zohra had had nothing to do with the plan.

Hamid himself felt uneasy, but he had no choice in the matter.

They had been six days at Mir Mahmood, when Zohra, taking a stroll in the garden, met Hamid returning from the city.

'You are late this evening,' she greeted him.

'There was a meeting of the Poetry Society, did you not remember? I wish you could have come.' They both started to walk towards the house.

'Was it interesting?'

'Yes,' said Hamid. 'Our poetess Sarojini Naidu was the surprise guest of the evening.'

'I did not know she was back in Hyderabad,' remarked Zohra. 'She is one of Gandhiji's staunchest lieutenants and so often travels around with him.'

'She has returned home to say goodbye to her friends and family,' Hamid explained, 'for she expects to be in prison again soon. The songs of "India's Nightingale" may be hushed under the burden of politics, but poetry is still her first love.'

'Yes, she is a most remarkable woman,' admitted Zohra. 'She has a soft spot for you too, hasn't she?' she asked as she looked archly at Hamid.

'You are jealous,' he remarked thoughtlessly, but at once a self-consciousness fell over them both. In an effort to get over it, she said:

'There was a message from Ammi Jan. Mehrunnissa has had twins.' She sounded thrilled at her sister's performance.

'*Mubarak*! That is exciting!' he exclaimed with his boyish smile which appeared whenever he was enthusiastic.

'It is the first pair of twins in the family. It must be nice having twins; saves such a lot of trouble,' she said.

'You mean, it is nature's labour-saving device for women who desire large families?' He gave an amused questioning glance, but she evaded it. She remembered how she had longed for twins when Shahedah was born.

'But I do not think Mehrunnissa will be pleased. She already has three—these two make five children!'

'An error of judgement on nature's part!' observed Hamid and they both laughed.

But suddenly Zohra became quiet as she thought of her sister. Mehrunnissa's husband, a weak character, had started keeping mistresses in the house together with the children he had by them. Mehrunnissa, at first jealous and rebellious, had now fallen into a lethargy. She no longer delighted in clothes or ornaments. Her figure was wholly ruined. Her children, never well cared for, were now entirely neglected.

Hamid and Zohra went into the house. Hamid wondered why Zohra desired a large family. Anyway, it was good for the world if women like her had more children, he reflected as a queer feeling overcame him.

As the days went by and Bashir gained strength, he became more and more absorbed in work. He had few real friends who cared to come and see him.

Zohra's friends and relations visited rarely, as women did not like to travel away from the city without a male escort, even in private cars. Zohra was therefore often lonely. She read a great deal but even that she sometimes found tiring. She tried to paint, but somehow the mood did not possess her. She was often restless. Safia's attitude, too, continued to hurt her. She never once came to see them, nor did she send Iqbal with Shameem and Shahedah when they visited Mir Mahmood.

Hamid was trying to finish what he had been writing for some time now. He retired into the round pavilion, away from the house, to work. He knew the only way he could go on staying there was to isolate himself. He avoided the midday meal by having coffee and toast sent to him. Even Zohra dared not disturb him. His friends sometimes visited, bringing news of the bookshop. Whenever Khorshed called, he asked Zohra to join them. He now knew for certain that he could never marry Khorshed, and wanted to prevent any further misunderstanding. He felt a cad for the way he had once encouraged her. Zohra sensed that Khorshed resented her presence, and was reluctant to be with them. Yet, Hamid insisted, and she could not refuse. Altogether relations between the three of them became strained and awkward.

Zohra had named Hamid's pavilion the 'Hermit's Cell'. It was open on three sides and overlooked a clear lake and green vistas studded with blazing red and orange *gulmohor* trees.

Once, early in the evening, she strolled out into the garden, and was tempted to walk past his retreat. On seeing her, he called out: 'What are you doing?' His voice was friendly.

'What are *you* always doing here on your own?' she questioned in return. She strolled up and leaned against a wooden pillar.

Hamid had not yet confided to her the nature of his writing, and he hesitated before he replied:

'I am trying to decide the fate of two young people.'

'That sounds God-like,' she laughed.

'Come in and I will show you,' he said on an impulse. 'Perhaps you can help me out of the dilemma.'

Zohra went in. Hamid, offering her his own chair, moved towards a wooden bench, upon which were strewn his papers.

'What is it—a novel?' she asked, sitting down in front of the table.

'It is a kind of socio-political tale, but it's really more of an effort to rouse Muslim nationalism against communalism. But, as you know, a romantic interest is essential, especially to hold a woman's attention.' He tried to be flippant in order to conceal the depth of his thoughts.

'It sounds good; but then I am only a woman!' She smiled in response. 'But what is the setting for your story?'

'It starts in Hyderabad. Two young people, although very much in love, decide to postpone their marriage and go to join the Satyagraha movement. Their romance is thus disrupted for the time being.'

'But romances do not exist in Hyderabad,' she interrupted with an impatience wholly unnecessary to the occasion.

'What do you mean?' he asked, startled by her abruptness.

'Oh, you know ...' she faltered, then continued hastily, 'people get married first.' She wished she had not so impulsively launched on such a delicate subject. She did not now know how to proceed.

'And yet most marriages here are successful,' observed Hamid in an impersonal tone, trying to avoid sarcasm. 'But I was not writing of arranged marriages, in any case. One has really to understand the psychology of persons entering into such marriages before attempting to write about them.' There was provocation in his voice.

'Oh, girls are brought up to expect their parents to arrange marriages for them.' She spoke with an affected carelessness. 'It is seldom that any other ideas enter their minds.'

'It is amazing. Any girl and any boy, and yet they live happily ever after!' Hamid's voice was disciplined, sounding as if he were only in search of knowledge.

'Not any girl and any boy!' she cried, but her voice faltered and an undercurrent of tension was now quite apparent. 'The parents in their own way usually take great pains to find a suitable partner.'

Hamid's hands were busy sharpening a pencil with a pencil sharpener, which appeared to be blunt. Seeing her hesitate, he encouraged her with: 'Yes?'

'I can only speak for girls,' she said quickly. 'They are brought up to look upon marriage in the same way as they look upon birth and death. In none of these do they expect to have any voice.'

'But what happens if a girl rebels?'

'You should know that a girl openly rebelling is almost unheard of in our strict Hyderabadi society! Why, do you not know that marriage is a forbidden subject between girls and their parents?' She was getting more worked up than the discussion warranted.

'But, according to Islamic law, a girl's voluntary consent is essential,' he argued, on an irresistible impulse to probe her mind on the subject. 'She can withhold that.'

'Really, where have you been living all these years?' she chided him, as her voice rose to a rather high pitch. 'A Hyderabadi girl would no more think of questioning her parents' right to arrange her marriage than she would of questioning God's right to dispense birth and death.' The subject evoked from her this spontaneous outburst.

'So that God and parents between them can dispose of all the three fates?' he asked with a strange look. 'And a girl is ready to accept her parents' choice, and offer her husband love and loyalty as naturally as she bestows affection upon her God-chosen parents. Is that the idea?'

'You have summed up the case beautifully,' she answered, quietening down with an effort.

'But marriage is such a personal affair.' His eyebrows were knit close together now, as he tried to focus his attention on his pencil. 'Mutual attraction I should have thought was essential.' He was still avoiding her eyes.

'The purdah system is a great saviour,' she was roused to say. 'This mode of marriage can hold good only in a zenana society, where there are no standards of comparison. Difficulties would arise if the girls started moving about in mixed circles.'

Throughout the conversation she had tried to maintain an impersonal tone but it was obvious she was finding this difficult. 'But tell me more about your plot. How do you propose tackling your problems?' she asked hastily, trying to get away from her churning emotions.

Hamid gave her a brief outline of the story. As she became interested, he promised to let her read the whole of the yet incomplete manuscript.

'I am wondering whether to give it a sad ending or a happy one.'

'What is more usual in life?' she asked.

'Tragedy, or quasi-tragedy. Disillusionment.' He was trying to sound detached.

'But for that very reason, people might wish to read a different ending to a book.'

I have already changed it once. I started giving it a happy ending, then altered it to tragedy. Now, again, I am uncertain.' He laughed in deprecation of his own indecision.

'It had never struck me before how easy it was for a writer to play with the fates of his heroes and heroines,' she commented.

'Yes, in a few strokes one can change their destinies completely! It's fascinating.'

'I wonder if God also enjoys playing with our lives in this fashion,' she said, gazing in the direction of the lake.

'Sometimes I feel it is all a deep preconceived plan. At other times, I feel things just happen haphazardly as we go along. Anyway, it is not always an amusing game,' he said.

'Problems could be resolved more easily if only one were single-minded,' she observed, alluding to his indecision regarding the ending. 'But life tosses one this way and that with such swift jerks that it becomes difficult to keep one's balance.' She did not quite know why she made this remark; it was just forced out of her.

'Problems are like the hydra, as soon as we solve one, two more crop up in its place.' Hamid made a gesture that seemed to dismiss everything. A silence fell over them both.

Hamid had noticed that despite these beautiful surroundings, which should have fulfilled Zohra's intense appreciation of nature, she seemed lost and unsettled. Having marked with concern Safia's antagonism towards Zohra, he wondered how much of her present restlessness was due to this. Hamid had tried to speak to Safia about it but had only received offensive replies with dark implications, warning him against her spell. The ties between Safia and her once-beloved brother had become strained, and Hamid was even sorry that Zohra had given her son away to her.

At first, Bashir demanded all Zohra's attention but now much less so. Although Hamid realized that his own problems were insurmountable, he still thought it his duty to help Zohra and try to understand her needs.

'Zohra, why don't you do some sketching here?' he suggested.

'I did try,' she said, 'but somehow I can't seem to keep up the interest.'

'Will you illustrate this?' he asked impulsively, holding up his manuscript in an attempt to give her an incentive. But even as he said it he wondered if he had spoken wisely.

'I? Why, I find I cannot do a thing. I shall probably make a mess of it!' But the thought was exciting to her.

'I have seen some of your old sketches,' he said. 'You could do it if you tried.'

'Anyway, it is one way to keep me occupied. Afterwards you can make a blazing fire and burn them all or throw them into the waters there,' she said, pointing to the lake.

Hamid and Zohra discussed the project and a strange exhilaration came upon her at the mere thought of sharing something with Hamid. She started on the illustrations that very night, and soon became engrossed in them. But she did not seem to be able to focus on her ideas for long.

Bashir was pleased to see her thus occupied. He did not understand why she was so dispirited much of the time. Perhaps she missed the children, he thought. He now spent his mornings and late afternoons working in the garden to get the benefit of the sun when it was not too hot. One day, he noticed a small boat tied up by the water's edge. Both Zohra and Hamid were with him.

'Why don't you two go boating? Hamid, you can teach Zohra to row. It's really not good for her to be tied to a semi-invalid all the time,' he said. His voice held a genuine concern for his wife. Besides, he would be able to concentrate better if he knew that Zohra was occupied pleasantly. He himself worked best when he was alone.

21

A couple of days later, after Zohra had struggled to row the boat smoothly, she put down the oars and, looking at her palms unaccustomed to such exercise, rubbed them and said: 'Let us drift along in peace for a while.'

Hamid merely nodded.

The Deccan landscape, containing some of the most ancient rocks in the world, was spectacular. Scattered around the countryside were smooth black boulders of varying sizes, some as large as a house, often balanced one on top of another as delicately as the brass water pots on the heads of Maratha women.

In the dazzling sunlight, the bare rocks appeared dark and ominous, although shrubs and green grass that sprouted through the crevices gave them, in places, a mellowed look. Hamid, watching them, mused: 'Under the burning rays of the southern sun, how rapidly these rocks absorb and radiate heat, and at the close of day, they quickly cool down and everything seems fresh and inviting once more. How closely in this, do they mirror the moods of Hyderabad's ruling class. For, are not the emotional Muslims quick to flare up and as quick to calm down? Warm and exuberant in their hospitality and affection, they take offence easily and as easily cool off.'

Hamid briefly reflected on his life since his return from England, vaguely dissatisfied with what he had done or not done. But his thoughts became disjointed as a deeper and a more pressing current threatened to overpower him. All around he could see, in small clusters, green trees—*ber* and *sitaphal*, the latter with sweet and luscious fruit, named after Sita, the devoted wife of Rama, who on her long wanderings with her exiled husband, had found them delicious and sustaining. Also, here and there stood tall, spreading *gulmohor* trees with sprays of dazzling red and orange flowers that shimmered in the evening light.

Gradually, the sun started disappearing below the western horizon and the golden light cast a warm glow on their faces. The chirping of birds at sunset greeted their ears; they still glided along in complete silence till at last Zohra said: 'You are so absorbed in yourself.'

'So are you,' Hamid countered. 'Human beings are really solitary creatures; and their thoughts are the only faithful companions they have in this world. But I confess I sometimes feel I cannot cope with any of it.'

'You feel like escaping from everything and everyone—your friends, your admirers; whilst they are always at your beck and call. You are only ever lonely by choice.'

'What you mean is, I am not alone by compulsion, but I can be lonely all the same.' After a pause, he added, 'In the ancient days of leisure, philosophy held a much larger place in our lives. The present-day bustle is alien to the Indian temperament. We should once more cultivate the habit of contemplation.' Then casting a swift glance at her: 'If I had any say in the matter, I would introduce the study of Yoga into our educational system; it gives one both mental and physical poise. I myself would like to take it up more seriously.'

'This place is ideal for meditation,' remarked Zohra.

'Perhaps some day I shall retire alone to some such place, at least for a period.'

'You always want to run away from reality. What you really need is a wife, a soulmate!' She laughed, but her heart leapt strangely. Hurriedly she quoted Ghalib:

> *Impossible is it to pass through life without love;*
> *But thou hast not the strength to bear the sorrows of love.*

Suddenly, and for the first time, Hamid now fixed his intent gaze on her: 'Zohra!' his voice was heavy with the richness of love as yet unprofessed, 'don't you realize that I love you?'

Zohra froze, as if her breath had left her body.

By no means had she been unhappy with her life. Her husband was good to her, and her children gave her profound pleasure and satisfaction.

However, in a vague and undefined way she knew that she had been waiting for some elusive happening, which would fill an empty space hidden somewhere deep within her. Now all at once, she knew what it was. A door had opened in an instant to reveal a light, which touched her spirit and permeated her soul. Suddenly, sunlit with love, she glowed in every fibre of her being. Not knowing how to look him in the face, she closed her eyes, in wild confusion, and fidgeted nervously with her hands. Hamid, still watching her intently, said:

'Zohra.' There was disbelief in his voice. 'You care for me too, don't you?' She could only give a faint nod. Hamid became silent, tense. After what seemed to them a long, long time, with a jerk of his head, Hamid said: 'Let's go somewhere and talk.' Zohra quietly acquiesced.

Hamid took up the oars again. Gaining the opposite bank, he got down first and offered Zohra his hand, but she instinctively withdrew from his touch. An extraordinary feeling, as yet unfamiliar, stirred her entire being, warming her into sudden life, and she seemed to lose track of everything, save the giddiness that was beginning to overcome her. She jumped out of the boat unaided, her sari getting caught under her foot and tearing. Hamid tied up the boat and came to Zohra.

'Let us sit there,' he said, pointing to one of the smooth black boulders that stood like a platform, cool in the evening breeze. They sat down a little apart. Zohra, suddenly bashful, drew the *pallau* of her sari closely over her semi-bare arms and, tightly encircling her raised knees, lowered her head onto them to hide the emotions that had so suddenly overtaken her. The folds of her sari, draping her graceful figure, were trailing right over her feet, and, to Hamid, she appeared to be as unreal as a sylph, apart and withdrawn. Hamid stretched out his legs and leaned back with his hands resting on the ground. Each seemed to be in awe of the other.

At last he called, almost reverently: 'Zohra!'

Startled, she raised her head and turned towards him without looking at him.

'Zohra, how lovely you look! I could sit and gaze at you forever!' Hamid spoke haltingly. 'Zohra, why don't you speak? Say something!' But Zohra felt herself drowning in a silence she found impossible to dispel.

'Did you know that I loved you?' His eyes now reflected a shy reserve.

'Yes,' she murmured.

'I loved you from our very first meeting. Only I didn't want to admit it, even to myself. You remember your throwing that garland round my neck, and our eyes meeting in confusion—or was it recognition? I had never seen anyone more beautiful; and how embarrassed you looked ... Do you remember, Zohra?' He was smiling his fascinating smile, through which his white teeth gleamed.

'Can I forget it?' she murmured, her eyes half raised to his which she dared not quite meet.

'You reminded me of a daffodil; you wore a golden sari, but I think it was the way you held your head, with a proud but graceful bend. But, Zohra, when did you first realize that you cared for me?'

'I don't know. But, I feel I've loved you all my life,' she said hesitatingly. Her voice was like the whisper of water, and Hamid had to strain his ears to catch her words.

'But we had never even met until my return!' He burst out, amazed at her confession.

'I know, but ...' she faltered and could not proceed.

'But what?' he encouraged her with a smile. He wanted to take her hand in his, but controlled himself, for he did not know what her reaction would be.

'In my heart I had idealized love and romance,' at last she forced herself to say in a low tremulous voice. 'Only, I thought they were just a poet's dream. When you arrived ...' she hesitated not knowing how to express herself.

'I filled that place!' He laughed in amazement. 'This is all very flattering. Only, don't idealize me. I have more than my share of human weaknesses.' He spoke with his eyes focussed on some faraway object. 'My only virtue is that I love you as I have never loved before, nor had ever expected to love.'

Their eyes met for the first time, and held each other as if in an embrace. They could not speak. It was Hamid who first broke the silence. They talked, engrossed in each other, secluded from the rest of the world.

It was growing dark; Hamid was the first to realize it. 'It is late, and we must get back. Brother will be waiting.'

It was their first encounter with reality that evening. They felt no guilt at their love, but they were both aware of the difficulty in having to face Bashir, for they could no longer talk as they had done before their feelings for each other had been revealed. Hamid also felt a pang of jealousy, which he tried to suppress.

The next morning Hamid retired to his shed very early without greeting Zohra, but everything he tried to write was surcharged with her: Zohra, Zohra, Zohra, ran through his mind. How would she like it? When would she read it? How would she look? His eyes constantly wandered towards the house, trying to catch a glimpse of her. He could do little work with his mind so obsessed, but he was determined not to go into the house.

Zohra tried to paint but, with every stroke, it was Hamid smiling, Hamid talking, Hamid lingering beside her. How could she express anything else? Her mind was so confused but at the same time she was filled with ecstasy.

The whole day they consciously avoided each other; but her eyes strayed towards the shed, even though she could not see Hamid.

That afternoon, Zohra selected a golden-coloured sari and dressed herself with special care; yet she was doubly careful not to do anything that would betray her either to her husband or to Hamid.

When the three sat together in the garden for tea, she could not meet Hamid's eyes, but watched him stealthily. She could hardly wait till Bashir had retired into the house and Hamid and she could once more be alone. There was the happiness of reunion. They walked to the boat and rowed out to the opposite bank. There, they sat on the same boulder as the day before.

'Are you glad, Zohra, that I spoke?' asked Hamid.

'What do you think?' She looked up at him. He could see plainly that there was no need for the question.

They recalled a number of past incidents, which had made them aware of each other. They were like carefree children, revelling in their newly acquired joy. At the moment they wanted nothing more from each other but the knowledge of each other's love.

'Why do people say it is wrong to love? How can an emotion that leads to such spiritual exaltation be wrong?' Her eyes were tender as she turned to him.

'There is nothing wrong in love, but you are married to my brother.' A sudden hopelessness came over him.

'But what are we doing that is wrong?' she faltered, not knowing how to express herself. 'That is ... marriage,' she at last dragged out of herself in a subdued tone, 'but this is love—the expression of one's deeper self. It cannot be acquired.' Her tone conveyed that the other thing could be procured. 'As Ghalib has said:

> *There is no compulsion in love, for it is that fire, O Ghalib,*
> *Which can neither be ignited, nor quenched at one's will.*

'My dear, all that is true, but love will demand other expressions—fulfilments.' He spoke guardedly, not knowing how she would take it. At that moment, she looked so uplifted that it seemed almost a sacrilege to hint to her of physical love.

'But ... but, we know that ... that is impossible.' She faltered, not daring to look at him.

'I know, and that is why I have suddenly started wondering whether we were right in igniting it. So long as we were silent, there was a veil of inhibitions between us.'

'Is it a sin to love? To feel the rapture of the heart?' she asked.

'There is no sin in loving. Personally, I feel even physical love when accompanied by an emotion as deep as ours is ennobling. It is the culmination of everything else. According to Sufi philosophy, it is through union with the beloved that one tries to reach oneness with the Divine. Love is the most beautiful thing, but what are we to do in our present circumstances?' He seemed tortured almost beyond endurance.

'But love is more a quality of the mind. It is the stirring of one's imagination; it is the fulfilment of one's spiritual self.' She looked exhilarated.

'Let us hope that it brings us happiness!' He sounded pessimistic. He had an overwhelming desire to touch her, to hold her hand, but he dared not.

'Let us make the best of what we can get—be happy in each other's company.' But even as Zohra tried to rationalize, she wondered if this would not be asking for the impossible.

'Zohra, you—a married woman—talk in this innocent manner?' Hamid sounded exasperated. 'You should know better!'

She could not reply. Her lips quivered.

'My dear, I am not sure we can go on with this,' he continued. 'My courage fails me. I cannot be with you alone, day after day, and not be tempted. I am afraid of myself. You remember once I told you not to awaken the sleeping devil in me.'

'Have I done wrong in loving you?' she queried.

'Sin only exists when there is consciousness of sin,' he answered. 'It is one's own approach to a situation that makes it right or wrong. Wine may be the nectar of the gods to one person, but poison to another. Ghalib, the poet, says: *What devil seeks happiness in wine? It is only a moment of forgetfulness that I crave, night and day!*'

'In my imagination I had placed love on the highest pedestal, as something unattainable. I had deified it.' She spoke with almost a devotional fervour.

'Yes, one may enter a shrine like a devotee, to worship, or like a sceptic, merely to admire the form and scoff at the spirit,' he commented sadly.

They were silent for a while. He then burst out almost in anger against himself: 'Oh, had I only returned from England immediately after my studies, instead of loitering around, seeking God knows what, probably we should have been married. Safia always talked about it. But it was not to be!' He gave a shrug of despair.

'Oh, why did you not come back!' Her voice came like a wail. 'As for my marriage,' she spoke hesitantly, as if struggling to have her most intimate thoughts reach out to him, 'I had a foreboding. You can never know what tears I shed on my wedding day. I had looked forward to love, where two spirits could commune with each other, and what I felt was the funeral of my dreams. But after the marriage, your brother was so good to me that I was contented, or at least so I thought until I met you. In a way, I even created an illusion of love for him, but now I know the difference. I now know why my heart ached and why I wept so bitterly that day.' Even as she said this, tears welled up in her eyes, for she knew that on that day she had forfeited her freedom, her right to choose.

Hamid, overflowing with love and compassion, gazed at her and under his gaze, she gave way and burst into violent sobs. Without pausing to think, he drew her head onto his shoulder, and though he was himself greatly disturbed, held it down gently. She was vaguely conscious of this tender gesture, and troubled by it. And she wept all the more bitterly. Gradually, as the sobs subsided, she lifted up her head in embarrassment and alarm. As she fumbled for a handkerchief, Hamid offered her his, then walked away. After a few minutes, feeling somewhat calmer, he came back to where she still sat in bewilderment, and lowering himself beside her, said:

'Zohra, today your head rested on my shoulder. It just happened. It was natural in the circumstances. Tomorrow something else may happen.' He spoke quietly. 'One thing leads to another. When two people are so greatly attracted, they have either to belong to each other wholly, or to part.'

She said nothing, but involuntarily caught hold of his sleeve, as if she could never let him go. They sat together a little longer. Suddenly, Hamid realized the time. 'It's late and the clouds are gathering. We have been too absorbed to notice.'

They hurried back. As they entered the sitting room, Bashir turned sharply in his chair to confront them. They had noticed that since his illness, he had become excitable and increasingly anxious. Addressing Zohra, he said: 'It is chilly, with the damp monsoon air blowing. You might easily have caught a cold. Sometimes you still behave like a young girl!'

'Perhaps I feel like one,' she answered, sitting down near the window and trying not to take his displeasure seriously.

Hamid felt a wild desire to shout at him, 'But why are you always after Zohra? Why can't you let her lead a normal life? Since when has she been susceptible to colds?' But with a sudden pang he realized he had no right over her, whilst Bashir, as her husband, had every right to be solicitous and worried. The thought was maddening. He stood trying to master his reactions, his knuckles white as he grasped the back of the chair.

'I know you love boating,' Bashir now continued in a more normal voice, 'but there should be a limit to everything. Look, it has already started raining; you might have got drenched! The lake is small. Where were you all this time? I couldn't see the boat. I wanted to signal to you.'

'Oh, we got down at the other end.' She spoke carelessly. 'We were just talking.'

'But what on earth were you talking about that you should forget the time and place—and the threatening clouds?'

They became self-conscious, wondering if he suspected anything. Hamid had already decided that it was for Zohra to tell her husband whatever she wanted to. He saw her pull herself up. There was a quiet determination about her, an honesty of purpose, that disliked taking shelter behind lies. It seemed as if she would tell him everything and face the consequences, if only he were to ask. She had no time to consider what would be the effects of her revelation upon Bashir, whether there was any kinder way. Hamid felt a wave of admiration for her courage rise within him; only he knew it would be foolish to wreck his brother's peace of mind as well as their own. What would they gain by it? Strange, he thought, how reticent she was when speaking of love, but faced with a situation where real resolution was needed, her spirit rose undaunted to meet any challenge.

Whilst the two were expecting an explosion from Bashir, the latter merely continued:

'Hamid has nothing to talk about these days save the Congress, Gandhi and his Satyagraha movement. He could talk all night long and forget the time. He does not realize that women do not have his khaddar sherwani to keep off the chill!' Turning to Hamid, he observed sarcastically: 'Yes, by wearing the coarsest homespun clothes and by submitting to violence in the sacred name of non-violence, you will indeed get your swaraj!' Then he added: 'Your faith in non-violence is quite nauseating!'

Hamid knew by now that any controversy on this topic invariably ended up in a collision with Bashir but, with an effort, he said: 'Last time it almost succeeded. It united the Hindus and Muslims as seldom before, at least since the British Raj. Why, even you were in agreement then.'

'But we have seen that it failed.'

'That was not a failure.' Hamid now turned towards his brother and retorted sharply. 'You cannot uproot anything so firmly embedded as Imperialism with one single wrench, but it has had its foundations badly shaken. It only requires one more effort. There is no other way. Non-violence is not only best suited to our conditions, but it is also the most civilized way. It is without doubt the highest form of courage, to suffer and not to hit back.' Hamid flung himself into a chair. 'Only the land of the Buddha could have produced the Mahatma. It requires the boldness of one's convictions; not a creed for the weak, it is a creed for the bravest.'

'We fight in the open battlefield,' said Bashir with cold finality. 'This method is certainly not suited to our Muslim genius.'

Hamid, losing all self-control, burst out: 'Why do we have to talk of the Muslim genius and the Hindu genius? After all, the majority of us come from the same stock. We are mostly converts, and have the same background of thought.' He banged the arm of the chair with his hand for emphasis.

'Even so, once converted, our social order has changed radically. In Islam, we have complete equality. With us, there's none of that inhuman caste system. You cannot shut up human beings in tight-fitting compartments in the name of religion. It is against the very spirit of Islam. The servant can rub shoulders with his master in the mosque. Before God, they are equal. By contrast, a Harijan may not enter a caste-Hindu temple. A Brahmin will not allow even the shadow of an untouchable to pollute him. What indeed is there in common between us?' Bashir asked in exasperation.

'I don't claim that Hinduism is free from abuses. But who is crusading more strongly against them than Mahatma Gandhi himself? And look at the tremendous effect of it even among women. Look at Zohra's friend, Nalini. She never ate with us before because her elders would not allow it. Now her family accepts it. Then, again, is Islam free from such abuses? Look at the way we have kept our women in the darkness of purdah. Is that human? Are we following the laws of the Prophet?' Hamid's voice rose as his temper spiralled.

'Certainly, there are abuses in Islam too. I'm all for reform, but you can't deny that Islamic Law, the Shariat, accords women more rights than any other religion. The laws of marriage, divorce, widow-remarriage, inheritance, they are all in favour of women. With the Hindus, on the contrary, child marriages and the plight of child widows are positively inhuman.'

Hamid squirmed, he felt like shouting: 'But how many of these rights you boast of are translated into practice? Can a girl really marry of her own free will, or get a divorce on any reasonable grounds?' But he restrained himself and said:

'Hindu philosophy is one of the world's oldest and greatest. One has to study it in order to understand the Hindu mind. It's no use attacking a system without an appreciation of its roots. And Hindu and Muslim cultures did reach a synthesis in the glorious days of Akbar. Look at Jawaharlal Nehru, isn't he the idol of everyone? Is there any Muslim to rival him?'

'Yes, handsome and impetuous, performing daredevil feats like a cinema star, he's at least the idol of all women,' commented Bashir icily. He was forgetting himself in the heat of the argument; for he had often confessed that Nehru was the only Hindu leader whom he could understand.

'I didn't know there was anything wrong in being idolized by women. I'm sure we should all like it,' said Hamid in an equally calculating voice. 'Only, Jawaharlal is respected by men and women alike.'

'This is just another form of idolatry,' broke in Bashir caustically. 'If you can't have a stone deity, you must have a clay one.'

'If that deity symbolizes some ideal, some perfection, then I for one see no harm in it. We're all doing it in one way or another,' returned Hamid, as the image of Zohra floated before his mind, even though his eyes were sedulously averted from her.

Bashir retorted: 'Muslim culture encompasses a graciousness of life, large-heartedness, tolerance.'

'Did Bashir imagine he possessed these qualities?' wondered Hamid impatiently. Aloud he remarked: 'Yes, Hindus don't squander money foolishly as we do. They have greater wisdom. I only wish, however, we would stop talking about different cultures. The differences are more provincial than religious. Religion is now being exploited only for power politics. The British used it in their policy of *divide and rule*. We were foolish enough to play blindly into their hands. And the way certain Muslim leaders are talking now, I dread to think of the future ... We *must* all unite and fight the battle for freedom. The youth of our country is mostly on our side.' Hamid's eyes were burning; his face quivering.

'Yes, you can always rely on the young to get excited over anything revolutionary. They are carried away by your slogans. It's just mass hysteria!' said Bashir. His monotonous voice was crisp and dry.

Leaning back more comfortably, he gave his brother a cold stare. It provoked Hamid still further: 'To me, Hindus and Muslims are one, possessing a common heritage, belonging to a common Motherland, fighting common battles,' he said. 'We are intermingled, living together side by side. Surely, you can't divide up each little plot of land and say this is Hindu India and that is Muslim India? To me that sounds like a graveyard!'

'I don't say that we have to divide the country up.' Bashir spoke in a hard metallic voice, as if dispensing judgement. 'I only say that there is justification for what some Muslims have started saying ... In a democracy we shall always be under the Hindus. We cannot tolerate that. We belong to the race of conquerors.'

'History has never shown a race of permanent conquerors,' said Hamid with an impatient gesture. 'Besides, Muslims can have their majority provinces with safeguards at the Centre. The British Raj can't last indefinitely. World opinion is already against them. And on the same basis, Hyderabadi Muslims have no right to rule the Hindu majority. We must all be prepared for changes.' Hamid spoke with intense conviction.

'The Hindus were not born to be rulers, they are yogis,' Bashir asserted with disdain. 'The Muslims never accepted British domination easily. Don't forget that we were the first to strike a blow at British rule. The Mutiny of 1857 was our attempt at freeing India. We fought an open battle, and not this ahimsa—non-violence. Even the Hindu diet is all wrong. No one can be strong on mere vegetables!'

'I don't agree with you there,' said Hamid with a jerk of his head. 'But anyway, in a free India, nobody is going to tie you down to sectarian habits. Even amongst Hindus, there are many non-vegetarians.'

'Altogether, it's their way of life against ours—the two can never meet. How can we come closer to each other when they won't even eat with us?'

'I am not defending Hindu conservatism, but I cannot defend Muslim narrowness either,' said Hamid heatedly.

'Muslims are not narrow,' Bashir was emphatic.

'That's our difficulty. Like the camel, we can never see our own hump.' Hamid gave a despairing shrug.

'You have no right to call yourself a Muslim!' Bashir's tone was so offensive that Zohra, no longer able to contain her anger, suddenly burst out:

'It is people like you who make one almost wish one were a Hindu. There should be some tolerance, a spirit of give and take. We all have Hindu friends, and what is the difference? Except, perhaps, that they are cleverer!'

'Hamid has been indoctrinating you. I do not think his influence is good for any susceptible young person.' Bashir's voice was scornful.

'A little while ago you asked me when I would stop acting like a young girl. I should now like to know when you will stop treating me like one,' she asked haughtily, rising from her chair. 'If I am susceptible, you cannot change me. I shall be influenced by whomsoever I like!' With this retort and her head held high she left the room. She had never talked to him like this before. Bashir, angry and hurt, and no longer able to bear his brother's company, followed her.

22

That night both Hamid and Zohra lay tossing in their beds, thinking of each other, thinking of their future ... thinking of the past ...

The next day, Hamid left early for the city, and Zohra could not set her mind to anything. Bashir was busy with his work, as if nothing had happened. But in the evening, after Hamid's return, wishing to make up for the previous day's outburst, he casually said:

'Why don't you go boating? Only Zohra, you should take a shawl with you and be careful of the weather.' He sounded solicitous, and Zohra felt a pang of remorse.

Hamid and Zohra still desired to be with each other. But within them the conflict was intense.

The boat glided quietly, the oars breaking the waters gently, and the calm lake readily making way for it to pass. All around was silence, as if the feelings of the lovers needed to be respected. It seemed that almost before the realization of their love had dawned, catastrophe was about to envelope them, sounding the death knell of their hopes and desires. Hamid's attempts at conversation sounded hollow and inept. They reached the opposite shore and Hamid said:

'Zohra, let us get out and sit and talk things over sensibly.' His voice was drained of all emotion.

They sat down again on the slightly sloping boulder. Hamid, with his elbows resting on his raised knees, was holding his temples between his hands. As Zohra looked at him anxiously, he said: 'We have dragged ourselves into a hopeless mess. I should have had more sense than to blurt out my feelings. I was a fool.'

'Is to love then, or to speak the truth, so foolish? Or is it that you find you were mistaken in your feelings?' She affected a light-heartedness in an attempt to drag him out of the depth of his musings.

'You know well that there is nothing in the world I would not give to make you mine,' he said, pulling himself up erect. 'But as that is impossible, I must leave you and go away.'

'Go away!' she exclaimed. 'But how shall I live without you?' The colour drained from her face and she began to tremble.

'But what is your idea of living with me?' he asked. 'For us to see each other and be tormented; for me to see my brother and feel resentful? Last night my eyes were opened to the things to come. Ordinarily, I could have warded off that discussion. For I know how unreasonable Bashir Bhai can be. But yesterday I was troubled, irritated, and why? Because he had the right to talk to you the way he did. Because he had the right to possess you.' Hamid's long fingers were now nervously tearing to bits some straw near him. He was getting so worked up, that Zohra quietly interrupted:

'He does not possess me. My mind never belonged to him—he had no use for it. Legally I am tied to him; but my soul and spirit are free. Both these are yours.' She spoke slowly, as if deliberating each word.

He regained some semblance of composure before continuing: 'You do not realize it now, Zohra, but you would be torn to pieces if I stayed. At present, I see no way out but to leave home. I shall pray for some solution. At the moment, unaccustomed as I am to prayer, I can think of nothing else.' He paused. His eyes were still fixed on his own busy hands. As she did not speak, he continued: 'During the past months, I have been trying to resolve where my duty as an Indian lies, whether even though I am a Hyderabadi, I should not go and join the Satyagraha movement. Then Brother fell ill and I had to put aside the idea, knowing that the movement could well do without me. Anyway, I knew I could join later; fresh recruits would be continually needed, as the earlier ones were shut up in gaol.' He gave a mocking laugh. 'Here, I was trying to finish that book before I could leave. But Brother no longer needs me. I shall now go at once and play the hero's part, knowing in my heart that it is cheap martyrdom, when the final decision has been forced on me by entirely personal reasons.' His voice was full of self-reproach. Zohra now understood why he had been sleeping on a bare wooden bed and eating coarse food.

'I cannot prevent you from going,' she said. 'In fact, I should like you to go ... but you will come back as soon as your work is done?' She was struggling to appear brave, while her heart was breaking at the thought of parting so soon after they had accepted each other's love.

'I do not know.' His tired voice seemed to be dragged out of him.

'You are not going away forever?' The words trembled on her lips, as the world around her seemed to grow dark.

'For ever? How can I know? In life, there is no such finality. All I know is that it would be impossible for us three to live under the same roof and keep our sanity.'

'Why?' she asked.

'How can I explain, my dear, if you do not understand it yourself?'

His tone was inexpressibly tender, knowing that he had no choice but to hurt her. 'If I loved you less, perhaps, I should have taken advantage of you, even though you were married. But this is hopeless ... Besides, he is my brother.' They sat there quietly, unable to draw apart, drained of all feeling save a gnawing sadness. They did not know how long they sat together, nor did they care what Bashir would think.

Hamid saw Zohra rise. She stepped down from the boulder, walked to the edge of the lake, and stood still and taut on a tuft of grass, her eyes transfixed upon the clear water below. Suddenly, fear gripped Hamid. He got up and swiftly walked towards her. As she heard his footsteps, she gave a start and would have fallen in, had he not caught hold of her and dragged her back to safety. Turning her to face him, and placing his hands on her shoulders, he asked fiercely: 'What were you doing?' For one brief moment, she looked at him with large sorrowful eyes. There was a confusion of emotions in that look. Holding her face up, Hamid impulsively moved his lips towards hers. But before they touched, he suddenly jerked himself away, dropped his hands, exclaiming violently:

'It is for this ... for this that I have to go away!'

She was speechless; her heart thumped as if it would leap into her mouth. He paced up and down for a while, then came back where she stood rooted, though quivering in every limb. He grasped her hand and led her back to the boulder. They sat down a little apart, side by side, she with her legs folded back at the side, the mauve sari covering even her feet. Hamid, his hands tightly clasped round his knees, turned to her, and said: 'I am sorry, Zohra. Can you forgive me?'

'What is there for me to forgive?' she asked gravely, sounding fatalistic.

'What were you trying to do? I was alarmed ...'

'I was looking at my reflection, trying to get my mind off myself. Oh, I am talking nonsense!' She made an impatient gesture with her hand, as if trying to clear away something in front of her eyes. The bangles on her wrist jangled harshly. 'And what if I had fallen in? To whom is my life of any value now?' she asked in a voice in which bitterness and sadness were mingled.

'I would have jumped in after you. We might have provided food for the fish!' He gave a mocking laugh.

'We should then, at least, have found a cool sanctuary. If we are not fated to live together, then we could at least have died together,' she remarked wearily, and sliding down a little on the sloping boulder she turned towards him.

He looked down at her, his face grief-stricken. What had he done to her? He blamed himself bitterly for having thus uprooted her from the life to which she had been safely anchored. He would readily have died if his death could have spared her this turmoil and unhappiness. He dared not speak.

'Men have other opportunities; *you* would not want to die,' she said, as if reading his thoughts.

'I? ... Yes, my life is so full! I am doing such wonderful things! What would the world do without me?' He uttered in total self-loathing. Then, to steady her emotions, now so exposed, he said: 'Think of your home, your children!'

'My home?' she asked, as if scarcely understanding him. 'It would only be a prison to me now. I was immured in it without even knowing all that was expected of me,' she observed very quietly. 'As for my children, what sort of mother would they have in a woman torn to pieces?' Her eyes flashed for a moment.

'Don't,' he cried imperatively, 'Zohra, you must not talk like this. You will get over it.'

'Get over this?' she burst out passionately. 'I also have tried to live up to certain ideals.' Her lips trembled. 'Now living with your brother would be right in the eyes of the world, but inside me I shall feel sinful, degraded. There is the sanctity of love which should not be desecrated.' Her voice was low but it vibrated with a quiet fervour.

'Were we in another country, under different social conditions, perhaps, some solution would have been possible. But here ... in spite of all our talk about Islamic rights, a woman's place is considered to be with her husband, however destructive that may be for her. Besides, his being my brother complicates matters all the more.' Hamid sounded wretched.

'I do not want marriage,' she forced herself to say, 'if I cannot live with the man I love, I would rather live alone.'

'You cannot do that,' he said with brusque finality. 'I implore you, think of the children, think of your parents, think even of your husband.' The last words were drawn out with a sharp pang of anguish. 'So many are dependent on you for happiness. You do not want to damage them for life?'

It was strange that Hamid should be pleading with her to stay with his brother, whilst, with all his heart and soul and being, he wanted her for himself. It showed a certain selflessness that Zohra could not help admiring. Strengthened by it, she said:

'You mean that one has to destroy oneself and all one's beliefs so that others might live?' she asked, looking up at him, as if he alone had the right to answer.

'Don't put it that way, my dear. You too will live, through them, and find your reward in their happiness; you will not be destroying your own ideals, but preserving the ideals of others.' He measured his words carefully in order to restore her composure.

'But how shall I live? What shall I do? If only you were always nearby to give me strength!' There was such a trusting look in her large liquid eyes, as she turned to him, that Hamid could only smile a faded smile.

'Child,' there was deep yearning in his voice, 'I have not the Buddha's strength. Do you not realize that love will not be satisfied with mere poetic imaginings? It is my fate that I must shun what I most desire ... How can I avoid making love to you when you sit there looking at me in that manner?' Words failed him as he looked at Zohra, the mauve sari draping her body in clinging folds, and spreading around her.

'As if you were my God, my lord and master ... Why do you not say it?' she asked so softly that the extravagant words sounded natural. 'Yes, you are all that and more.' She drew in a deep breath, and clasped her knees closer, the flowing sari folds being tucked away with this movement, like the petals of the morning glory closing up, in the glare of the noonday sun.

Her words moved him deeply, but he made an effort to laugh them off: 'Do not be absurd. It is not like you.'

'I have not seen God, but I have seen you.' she replied, looking at him with such reverence that he was left speechless; then, closing her eyes, she inclined her head.

After a time, she turned to face him again and spoke: 'There must be either something wrong with our moral code or with the laws of nature.'

Hamid felt like crying out. 'There is everything wrong with our society.' But he only closed his lips more firmly, determined to instil in her the strength to go on. He knew it was for him now to give her the courage to stabilize her life. He felt his heart against his chest overflowing with love for Zohra.

A silence fell over them as the awareness of parting filled their minds to the exclusion of all else. Hamid planned to leave Mir Mahmood next morning. His brother no longer really needed him. At last, Zohra, awakening from a long reverie, asked:

'What time will you leave?'

'Before breakfast,' said Hamid. 'I shall not see you alone again.'

Her lips quivered. Tears started to gather in her eyes.

23

The next morning, Hamid left Mir Mahmood. Zohra looked so ill with her nervous headaches that, after a couple of days, Bashir decided that they too should return. Zohra went to stay with her parents. Her family, as well as Bashir, believed that the anxieties and worries, caused by his long illness and convalescence, were now telling on her.

But her mother could not understand it; perplexed, she remarked: '*Owi*, Zohra, what has happened to you? Allah forbid, but someone who did not know how fortunate you are in your bridegroom would think that you were burning yourself out. *Ai-hai*, look at your face! The colour of ashes!'

'May Allah preserve the couple for ever and ever!' said Unnie fervently. 'May Allah preserve your bridegroom from the evil eye, but how anxious he is; he comes almost daily to enquire after you!'

Zohra listened to all this in silence. She alone knew what the matter was. Deep inside, she was being devoured by pangs of longing, and she could speak to no one. Nalini came often. Zohra felt that her friend, with her quiet wisdom, gauged the situation, yet never touched upon it. Instead, they talked of mysticism and philosophy, and of Gandhiji.

'The very fact that, as a Hindu, I can now stay here and eat with you all is the greatest proof of Mahatmaji's influence. My elders have ceased to think of it as a sin,' said Nalini. She herself was now married and had a daughter but Zohra could never make out whether she was happy or not. She had the look of one resigned to her fate.

Hamid left for Bombay soon after his return from Mir Mahmood. The bookshop was to a certain extent serving its purpose. Hamid had succeeded in arousing the interest of young Hyderabadis. They read progressive literature and were made more conscious of the problems confronting their country. Its management was now in the hands of Khorshed and Shareef, two of his most trusted friends. In fact, the business brain behind it had always been Khorshed's, who in spite of her loud and flaunting ways, had a very sane head. It was she, from the outset, who had recognized Hamid's shortcomings as a businessman. His long Nawab ancestry had made him entirely unsuited to such work, whilst with Khorshed it was just the reverse. Parsis were noted for their business acumen and organizing power. However deep may have been her disappointment at the absence of any emotional response from Hamid, she tried to hide it, and remained his supporter. But she guessed Zohra was somehow responsible for his sudden departure and she resented her.

It was the year 1930. News came of Hamid. He had broken the Salt Laws. This had now come to represent the destruction of alien power. The levy on salt by the rulers was a detested tax, for peasants looked on salt as a gift of nature. In March, Gandhiji began his march to Dandi Beach—a distance of 240 miles from Ahmedabad—with a group of his followers. There, in a symbolic gesture, he picked up a handful of salt from the dunes. This was a signal for millions of Indians within reach of the sea to do the same. Thousands were beaten with lathis and injured. Many thousands more were imprisoned. It was said that criminals were released from gaol to make room for political prisoners. Hamid also had been arrested and sentenced to two years, a very long term for such a trivial offence.

When news reached Hyderabad of Hamid's arrest, his friends accepted it with mixed feelings. They grumbled, cursed, felt strangely elated, and celebrated the occasion. But Bashir and his mother were hurt in different ways. Masuma Begum was all solicitude.

'What will he eat? I hear gaol food is not fit even for animals. Can we not send him something? *Ai-hai*, and how will he sleep? He was fastidious about so many things.' Tears melted her austere face as she thought of Hamid's predicament. In her deep concern, she forgot her anger against him for going away to join the movement.

'He will be all right,' said Bashir in a quiet but derisory tone. 'After all, he is not a child: he knew perfectly well what to expect.'

A letter addressed to all the family—Zohra knew she too was included—came from Hamid, written from gaol. He sounded cheerful: 'It is good discipline for one's soul. In detachment, one has time to think over things. There is a complete reorientation of one's sense of values.' He then enquired after everybody and casually asked, 'How are Zohra and the children?' Zohra read the letter again and again, in solitude, with trembling hands and an aching heart, trying to read hidden meanings, concealed messages.

There was also a letter from one of his co-workers, who gave news of Hamid:

'... within days of his arrival, Hamid Bhai has become the undisputed leader of our group. He commands the greatest respect from his team. When he is working, he is like a man possessed, organizing, writing, often through the night. His speeches are passionate and inspiring, much to the dismay of the powers of law and order, who have soon discovered that Hamid is a dangerous man. Yesterday, we sent Hamid Bhai off to Thana gaol with garlands made of homespun yarn, as is our custom ...'

That Hamid had been quite seriously injured in a lathi charge was not revealed to his family at his own request.

Zohra, reading this, thought Hamid at least had done something, achieved something, while she ...? But, she was only a woman. She would have gone willingly, leaving the children in the charge of their grandmother, for the duration, but she dared not even breathe a word about it to her husband. He had been furious enough with his brother.

Bashir waited for Zohra's health to improve and longed to have her back with him again. But she neither looked better nor showed any inclination to return home. At last, one day he said:

'Do you not think, Zohra, we should do something about you? What is the use of your staying on at your parents' if you do not improve at all?'

'What do you want me to do?' she asked, for she had never wished to hurt him.

'I have been thinking, and I am certain a sea voyage would do you good. Let us go to France. I am entitled to six months' leave and I have been wanting to do some research work in Paris at the Sorbonne. One gets fossilized here. I can apply for leave at once.'

'Yes,' said Zohra with more enthusiasm than she had shown for anything during the past few months, 'let's go!'

All of a sudden, she felt that if she had to resume her life with him, it was best to get away, at least for some time, from their home, from Hyderabad, from India; the further away the better.

Bashir was pleased. Zohra knew she could leave the children with Masuma Begum who would not spoil them like her own mother would. As for Hamid, he was beyond her reach, and there was nothing she could do for him. Besides, it was he himself who had wanted her to continue her life with his brother.

There was one more letter from Hamid before they left. In this he described gaol life almost with relish and gave no hint of any discomfort caused by the appalling conditions that everyone knew prevailed in prison. He talked of his companions; discussed how the trip to Europe would benefit Zohra in every way. He mentioned the places she should visit. Zohra yearned for him and wondered if it contained a hidden note of disappointment. She remembered their trip to Ajanta more vividly.

Bashir and Zohra travelled by an Italian liner, the *S.S. Conte Rosso*. Bashir had forced himself into a holiday mood as even a holiday was a cut-and-dried experience, which had to follow a prescribed routine for him to enjoy himself. He went out of his way to be sociable. He played bridge and exchanged drinks with his fellow passengers. Zohra sometimes joined him at deck-games. But she had almost immediately drifted into another group that collected in the piano room and held musical evenings. When diverse people are confined together, friendships are easily formed. Two Italian tenors sang operatic arias. Zohra listened intently for, to her, their voices were more beautiful than anything she had heard before. But it only enhanced the longing to share this experience with Hamid.

In this group was also a Frenchman, an artist, Jacques Verneuil, whom she found intriguing. He was tall and gaunt, with a long face set above a crane-like neck. His retroussé nose lent him a sort of unapproachable air. But, as if to make up for it, he had humorous grey-blue eyes. They made Zohra feel he knew something of what was happening to her, for they often held a hint of pity as they rested on her. A large jade ring on the little finger of his left hand had caught her attention. He had told her he had bought it from some strange dealer in Hong Kong, and that he was superstitious about taking it off.

One evening, Zohra stood on the deck, leaning against the railing and watching the sunset with unguarded anguish. These days, beauty caused her heartache and only helped to heighten the sadness of unfulfilled dreams. She was lonelier than ever. Jacques came and stood beside her. She turned towards him, almost with unseeing eyes; but immediately recognition dawned and she gave him a welcoming smile. Jacques marked the changing expressions carefully.

'Pardon, but Madame is lonely?' The voice was so sympathetic that without thinking she replied:

'But who is there not lonely in this world, Monsieur?' She tried to sound nonchalant but without much success.

'*Mais oui* Madame, but excuse my saying so, Madame's eyes ... they are unhappy.'

Nobody as yet had noticed her suffering. Suddenly, fearful of having her secret revealed, she said hurriedly:

'But, are we not all troubled in some way or another?' She laughed, hoping he would not try to probe too deeply. But he continued:

'I knew it, Madame, ... I knew it.' His eyes were benign. 'The trouble is there ... deep down, is it not?' His hand went to his heart in half-mockery, as if it were rather an unnecessary organ that one could easily do away with. Her startled eyes rested on his hand and the jade ring on the little finger.

As Zohra watched it, he continued: 'I knew it from the moment I saw you. Nothing is hidden from me.'

Zohra wondered if he had really been able to read her innermost thoughts in so short a time. She turned away to look out onto the sea, but Jacques was watching her intently.

'Madame, it is the eyes that give you away; beautiful eyes they are— deep, and they hide many thoughts.'

'Do they?' Zohra tried to laugh away her own guilty feeling.

'And expressive too,' said Jacques. 'I know that look. Maybe I can help, if you will let me.'

All of a sudden, Zohra felt an impulse to confide in him. Perhaps he could really help. He looked so understanding, and such things were not rare in his country. Somehow it would be easier to speak to a foreigner, a stranger. She could certainly not talk to anyone in Hyderabad. At last, putting aside her inhibitions with some difficulty, she started to speak and he was able to draw her out slowly, with kindness and yet with humour. When she had finished, he responded with a flourish of his hand:

'The only thing for you is to forget him.'

'I wish I could,' she said, trying to sound light-hearted, but her voice verged on despair.

'Why, Madame is very beautiful, attractive. She will have people fluttering round her like *papillons*, only if she would let them. Everyone is asking who Madame is. They think you are a great princess, a maharani of India. You know, you are so distinguished, so elegant.' He had been illustrating his sentences with expressive gestures.

Zohra turned her head away in confusion.

'Please, Monsieur,' she said, 'if you do not stop, I shall really start believing that there is no one in the world like me!' She was leaning against the railing.

'But there is not, certainly; Madame does not know her own charm. Wait, you will see in Paris how they fall for you!'

'And what will I do then?' she asked, feeling more light-headed than she had done for a long time.

'Play, talk, dance, love.' The last word was uttered caressingly with a spark of mischief in his eyes.

'Love?' she asked perplexed at what to her was heresy.

'Yes, Madame must know that to fall out of love, one must fall *in* love. Not so hard, but just like a little injection ... like a serum.'

She smiled as if it were a joke, but a joke that also hurt. Sensing it, he tried to divert her.

'And Madame must dance. On the deck, it is very lovely in the moonlight.'

'But I do not dance!' she exclaimed.

'Ah, but I can teach you. For a lady like you, it is not difficult. You need only somebody to guide you. Madame has a most graceful figure.' He made a rhythmic sweep with his hand. 'It is a poem, a real poem. What would not a Frenchwoman give and men too,' his eyes twinkled, 'to possess such a figure!'

Zohra felt ill at ease, but he appeared so unconscious of the effect of his words, she did not know what to make of it; she remained silent.

Every night an orchestra played in the ballroom, and people danced and enjoyed themselves, allowing a feeling of abandonment to creep into their comportment. That evening, Jacques suggested they join in. To her astonishment, Bashir said:

'Yes, Zohra, why don't you? There is no harm in it. It is good exercise, besides being a social accomplishment. In Europe it helps.'

'Bravo, Monsieur. Persuade Madame it is also soothing for the nerves.'

Tired of the lingering ache in her heart, she said:

'Very well ... but ...'

'Ah, Madame, it is excellent.'

'But you did not let me finish ... I shall only dance if my husband does!' She turned to Bashir, half expecting him to say 'no'.

'Yes, I think I shall take it up again. It is useful in Europe. Otherwise, one feels rather out of it.'

Bashir, no easy social mixer, always had an eye on what was expected of him in society. 'Why can't he do things because it pleases him to?' thought Zohra, a little piqued. But to Bashir she only said:

'Then I shall try dancing with you first. I shall see what it feels like. May I, Monsieur?'

'Certainly Madame,' but he seemed slightly put out. He could not quite understand her inner difficulties, the inhibitions that she was trying to overcome.

Bashir and Zohra went onto the floor. He tried to teach her the steps, but he was naturally clumsy and out of practice. Saying that he had had enough for the evening, he handed her over to Jacques. As Jacques swung her onto the floor, Zohra sensed a new excitement, a different rhythm as he led her with ease, and she soon began to keep in step.

'Ah, Madame is dance personified!' enthused Jacques.

Zohra did not feel very comfortable; it was too new an experience. Besides, as Jacques held her, it was only Hamid's hands, Hamid's face and Hamid's eyes haunting her, and instead of pleasure, it brought her greater misery.

When they returned to their table, Bashir ordered champagne. The waiter, placing glasses before them, started to pour out first for Zohra.

'No, not for me, please,' she stopped him with a gesture.

'It does not matter, Zohra; put away your scruples. We are no longer in Hyderabad. Take a sip and see; it is good. It will lift your spirits.' Bashir spoke as if this were also one of the social niceties he would like her to acquire. In Hyderabad, he had never asked her even to taste an alcoholic drink.

Ordinarily, no amount of persuasion could have induced Zohra to touch it; but since the awakening of her love for someone who was not her husband, her old moral code had been badly shaken. The rebel in her now made her think, 'Yes, why not try it? Perhaps, this will make me forget the anguish.' Therefore, to Bashir's surprise, for he had never expected her to give in so easily, she said, 'very well,' and quietly lifted up her glass. Jacques, raising his towards her, said:

'*A votre sante* Madame ... and to your future!'

His eyes held an expression, which she could not decipher. She took a sip or two and then put down the glass.

From that day on, two of her greatest inhibitions were broken. It was Bashir who had got her started. She soon picked up enough steps to dance, under Jacques's expert guidance. She also learnt to drink a little, and found that it had the effect of mitigating her obsession, at least for the moment.

There was a fancy-dress dance on deck before they reached Trieste. Zohra was reluctant to dress up for it but Bashir, aided by Jacques, insisted that she should.

Her resistance was low. Besides, she was in the mood to try out something new and exciting, for a degree of daring and defiance growing within her seemed to alleviate some of the pain.

Bashir had left her alone in their cabin, to let her dress in comfort. She put on a crimson and gold sari in the Mahratta style with a trailing *kashta*, as the maharanis wore it. Her old Hyderabadi ornaments set off the costume beautifully. She wore a chaplet of artificial flowers round the knot of her hair at the back. A crimson *tikka* on her forehead, like a beauty spot, brightened up her face.

As she gazed in the mirror, she experienced an excitement run through her, which was very satisfying. When Bashir returned, he gasped with admiration, although he was not usually given to displaying his emotions.

'Are you out to break hearts?' he asked, and came over to kiss her.

'Don't!' she exclaimed, pushing him away. 'My colour is artificial.' Bitterly she felt like adding, 'Like myself these days.'

He was aggrieved and moved away.

She entered the lounge hesitantly with her husband. Her friends immediately gathered round her with ecstatic exclamations, for Continentals, she had noticed, were generous in their praise. Never, except on her wedding day, had she been on such public display. But on that occasion there had mainly been women, whilst here the majority of them were men.

'Ah, Madame! Madame is a vision of beauty ... Like a nymph!' exclaimed Jacques with a gesture conveying his appreciation.

Zohra did look nymph-like that night, with the trailing sari clinging to her graceful figure. She won the prize for the most beautiful costume. But exotic as the costume was, it was evident that it was the wearer's own loveliness that enhanced its effect, and her admirers did not conceal that fact.

Before the party started to break up, Zohra slipped away alone to her cabin and, locking the door from inside, sat down before the mirror. She gazed at her own image with mingled admiration and disgust. Hamid had never left her thoughts throughout those long hours, and yet she had felt a sense of gratification at the flattery lavished on her that evening.

It is only when a woman has nothing else to live for that she wants to live on vanity, she bitterly thought to herself. 'There was not a single conscious moment when I could forget him even though I was so universally admired. Some had even reproach in their eyes for my apparent coldness. But of what use are they all? If only I could myself feel and make my presence felt in that dingy little gaol-cell!' Her eyes were trying to pierce through the prison walls. 'What shall I do with this?' And she looked sadly at a pretty bottle of French perfume called *L'Amour* that she had won for her prize.

24

They spent the first month in England, where Bashir visited his old college—Trinity—at Cambridge. They met some of his old professors and friends who were now dons or lecturers. He took a secret pride in introducing his wife, who moved amongst them with a reserved but distinguished air. They were entertained and taken to all the places they wished to visit. Zohra did everything that she was expected to, with refinement and style, winning hearts easily, whilst her own was dead to real happiness. She tried hard to put Hamid out of her mind, but the more consciously she tried, the more she missed him.

Afterwards they did some hurried sightseeing on the Continent, visiting Switzerland, Italy, and Germany. Then finally they settled down to spend four months in Paris, where Bashir had arranged to do some research at the Sorbonne. Bashir soon became preoccupied with his work and was relieved to have Jacques, their fellow-passenger on the ship, who was now in Paris take charge of Zohra's debut in the Parisian world.

Jacques immediately arranged for Zohra to study art under his own master, who had a studio in Paris. He himself worked there at times. One day he showed her a picture he had painted, which he said was inspired by her. It was of a figure floating through opalescent clouds, ethereal, mystic, and unapproachable. Yet the eyes revealed something that drew it unquestionably to earth!

'This is you—right?' he asked.

'I do not know,' she laughed uncertainly.

He told her, effusively, how he could not sleep at all, until he had got it off his mind and put it on canvas. Jacques also introduced her to some young people whose company he thought she would enjoy. They often sat in cafes, having drinks or coffee and talking and watching people walk past in the street. Zohra found this fascinating, although her knowledge of French was very limited. But often the others spoke English. And so she found herself in a group of people, suave, original, artistic, amusing, and full of creative vitality. They were not shackled by inhibitions. They, in their turn, were captivated by her looks, her naivety, and her shy aloofness. With them she often visited art galleries and exhibitions. She also went to theatres, operas, ballets, and to dances and soirées. Bashir usually returned from work late and tired, and seldom felt inclined to go out, especially to the places where she wanted to go. But he was anxious that she should see everything and enjoy herself.

Soon after her entry into Jacques's circle of friends, Zohra was persuaded to go to a beauty salon, where she acquired the subtler ways of make-up. She now had the appearance of a finished product, polished and sophisticated. One night she went to the theatre with Jacques and another couple, also artists, who were friends of his. Zohra was quite taken aback when she learnt that, although unmarried, they were living together and made no secret of it. At the same time she felt a deep envy. What would her own life have been like, she wondered, if she had lived in a liberal society? The sense of injustice brought tears to her eyes. Afterwards they all went to the *Beaux Arts* Ball. As she floated across the highly polished floor with Jacques, he said:

'Zohra, you are ravishing tonight. *Quel parfum!*' He inhaled it deeply. 'You are here to kill?' A gallant smile spread over his face. His eyes were amorous as he drew her closer to him. A cold shiver ran down her spine, transmitting the feeling to her partner. 'What is it you are always fearing? There is nothing wrong in love!' He clung closer, as their bodies moved in rhythm.

'Let us sit down,' she said, trying to extricate herself from his clasp even before the dance ended.

'As you wish.' He was offended but, trying not to show it, ordered the drinks. Apparently he had already drunk more than was good for him. But Zohra had no power to check him. Besides, she felt she herself was in need of wine to pull herself together. Hamid was swimming before her eyes. Jacques, although attracted by her, had until now always been considerate, trying to appear both interested and detached. But tonight his senses seemed to carry him away. She asked him if they could leave and he readily consented. She looked around for their friends but they were nowhere to be found. Frightened and nervous, she had no alternative but to get into the taxi with Jacques. Inside, his arm went around her, holding her close to him and, as she struggled to free herself, he ordered the chauffeur to drive down the Bois de Boulogne, where he asked him to stop. The driver discreetly got down and left them to themselves. Jacques now kissed her wildly, feverishly. His hands started to caress her. The more desperately she tried to push him away, the more tightly he clung to her. She had heard of women shrieking, but how could she debase herself by calling the attention of strangers to the scene? She could not think. She was merely acting on intuition.

'Please, please, let me go!' she cried, numb with fright. For a moment he held her away from him and, trying to gaze at her in that dim light with burning eyes, said:

'You must know what it is to love. You are a dead woman; I can make you alive!' His words came breathlessly.

'You cannot make me alive!' Her voice hardened. 'I belong to one; I love another; I will not have a third make love to me.' There was something so decisively cold in her utterance, that Jacques's ardour was instantly chilled. The words fell like icy water on his kindled emotions. He pushed her away. They found themselves at opposite ends of the seat. Then, summoning the chauffeur, Jacques ordered him to drive back to her flat near the Luxembourg Gardens. Whilst leaving her at the door, he formally kissed her hand and asked, 'We are friends?'

'Yes, for the sake of all that you did for me before; but we cannot meet again.' Her voice was flat, lifeless.

'This is farewell?'

'It has to be,' she replied in a strange voice while closing the door on him. Then she ran up the stairs.

As she entered their living room, she saw Bashir still sitting up reading. He looked up and noticed her dishevelled hair and her agitation. 'What's the matter, Zohra?' he asked sharply.

'What does it matter to you?' she flung back violently.

'What do you mean?' Bashir rose. He seized her by the shoulders and tried to look into her eyes. Another face swam before her, but trying to drive it away, she said:

'I mean,' her breath was tinged with wine, and her eyes, slightly glazed, had a hunted look, 'I mean, you knew the world and you left me alone to its mercy. You knew human nature and you did not care what I did, so long as you were left in peace with your books.'

He stepped back and, holding on to the chair, turned his stupefied gaze on her.

'Stop!' he said peremptorily. 'You are crazy! You know perfectly well that I left you alone because you liked being with them. You said you did not want to be treated like a schoolgirl; you thought you were quite able to take care of yourself.' Bashir's relief at seeing Zohra safe turned to resentment. 'You have had more drink than you can stand.'

'It was you who wanted me to learn all the social graces. It pleased your vanity. So here I am! You should be proud of me now!' There was dark reproach in her eyes, as she stood defiant.

'But what *has* happened? For God's sake, out with it, and don't stand there staring at me in that stupid manner!' Bashir was for once wholly upset.

'Why should I?' She gave a careless shrug.

'You're not yourself!' he burst out angrily and, again taking hold of her by the shoulders, shook her violently. He had never behaved like this before. The possessive instinct in him rose to the surface and with it his usual level temper vanished. Zohra wrenched herself free, almost shouting in her fury, 'Let me go!' She flung herself down on the sofa and, leaning against the cushions, covered her eyes with her arm, as if to shut out everything. Regaining his equilibrium, Bashir now tried to reason with her, but she snubbed him. 'I don't want to talk to you ... Why don't you go to bed?'

'But aren't you coming?' he asked.

'No,' she replied tersely.

Bashir hovered around, wishing to make up to her in some way; for already he was beginning to reproach himself for losing his temper. But noticing her look of complete indifference, he thought it best to leave her alone and went into the adjoining bedroom. It was impossible for him, however, to sleep or even to lie down. Calming down somewhat, Zohra went into the dressing room to undress. Bolting the door from inside, she threw off her wrap and sat down at her dressing table. She had of late got into the habit of scrutinizing herself. As she gazed at herself in the mirror, she felt she was looking at a stranger. She saw her mouth tremble; she felt she was now tainted. With a shudder, she wished she could wipe off the pressure of those wild lips that had pressed against hers. She looked at the sari-draped figure, the beauty of which she had become increasingly aware, especially since coming to Paris. But now a shiver of distaste ran through her. She unclasped her necklace with a violent jerk, as if wanting to cast off every trace of allurement. In disgust, she wondered what pleasure anyone could gain from such forced kisses and caresses. Suddenly, the trend of her thoughts changed. 'But why have I been dressing in this tantalizing fashion?' she asked herself. 'I have had more admiration and adulation than ever before in my life, and from so many ... and yet I have never been more unhappy. I was a novelty to them; I often saw desire in their eyes and took no heed. I wanted their acclamation and still expected to keep them always at a distance. Perhaps I acted in a way that tempted Jacques.' She gave a tired shrug. 'Why put all the blame on him? He had been a wonderful friend, shrewd but compassionate. It was he who helped to save my sanity. He was the only person I could talk to about my difficulties, and now I can never see him again.' But how indeed could she explain all this to her husband?

She gave a start, for she had an overpowering feeling that Hamid was near her, watching over her, protecting her with his wisdom. 'I knew you would understand,' she started to say to herself as if addressing him. 'What wouldn't I give to be near you! How willingly would I exchange all these glamorous clothes for a plain khaddar sari, and instead of these lovely surroundings, stay in a dingy little room in the sweltering heat of Bombay! To catch even a glimpse of you would be life again! ... I can stand no more of this!' She rose impatiently and, unfastening her bracelets, flung them into her jewellery box. With a turn of her body, she exclaimed with yearning: 'Oh, I want to go back, I must go back, although I can't even see you!' Then again in her heart she thought, 'But I can see my children; that at least would be some hold on life.' Her heart swelled with an urgency to return home.

Taking her blankets to the sitting room she spent the rest of the night curled up on the sofa. Protests from Bashir had no effect. She slept in nightmarish snatches. Next morning, neither she nor Bashir alluded to the previous evening. But a curtain seemed to have fallen between them, and each behaved in a formal and civil manner to the other.

Before going to the laboratory, Bashir asked, 'What are you doing today?'

'I don't know.'

'Would you like to go out with me somewhere this evening?' he asked but she perceived a coolness in his attitude.

'If I am in the mood, yes.' The chill in his voice had provoked her into making this somewhat insolent reply.

Exasperated, he took a deep breath, and came and sat beside her on the sofa. With a visible effort he controlled himself and said: 'Why have you become so moody these days? The more freedom you have, the more restless you become. What do you want to do, Zohra?' He looked concerned now. 'You will have a breakdown if you go on like this. What *is* the matter?'

'I want to go back, I must go home,' she replied soberly.

'But I have not finished my work here.'

'I know, but you can stay on.'

'But how can you go back alone? Besides, I shall be lonely.' He was sincere, but she laughed a little cynically.

'You do not have much time in which to be lonely. Besides, this is Paris. If you want a woman's company, you ...'

'Stop. What is the matter with you? You never talked like this before. As if I would!' he said appalled.

'Oh, no, you are much too self-righteous!' Her voice was bitter.

'Please, Zohra, stop this nonsense, and tell me what you really wish to do.' He was struggling to keep calm.

'I have already told you. Please find me a nice old lady who is going to India to chaperon me on the voyage, and I shall go.' Her voice sounded strange, and Bashir could not say whether she was mocking him for his fears on her account, or was serious about the chaperon.

Ultimately, it was decided that she should return. Fortunately, she got a passage on a boat sailing a week later, and prepared to leave. She could never bring herself to tell her husband what had happened that night. For him to understand, she would have had to tell him so many other things. And, for his own sake, that was impossible. Bashir, realizing that she was definitely going back, felt drawn to her again. He noticed that Jacques, hitherto almost a daily visitor, had stopped calling altogether. He asked her once or twice about it but getting only vague replies, he kept his speculations to himself. But he found her more considerate towards himself.

The evening before she left, she said sadly but affectionately: 'Forgive me. You deserved a better wife.'

'But, Zohra, you never told me what happened. Why this sudden decision to go back? Could you not have stayed on?'

He wove his fingers through hers, as they sat on the sofa. Returning the pressure, she said: 'I am sorry, I have to go. There is nothing I can explain. I know I have been unbearable. But, believe me, I never wanted to cause you suffering.' She spoke in short, jerky sentences and could say no more, nor could he question her any further. Zohra returned home, sad but relieved to be back. Her disillusionment was great; for she had sought forgetfulness and had only found greater unhappiness. The things she had done, she thought, were perhaps suitable for those who were brought up to them, but she could not fit into that pattern, and remain carefree.

25

'How do you feel, Monsieur Vice-Chancellor?' asked Zohra with mock gravity.

'It is a wonderful opportunity. I can now reorganize the university and besides, it is a position worth having.' Bashir's somewhat lofty tone jarred on Zohra.

'I have rarely seen you so elated,' she said, trying to hide her disappointment in him beneath a faint smile. Bashir could seldom spare much time for Zohra, but today he said:

'Let us go to see a film. We can go to an out-of-town theatre where segregation of the zenana is not strict. Then we can sit together.'

'If you had told me beforehand, I could easily have managed it, but now ...'

'Yes, of course, I forgot your classes. Your widows' sewing class, and your orphans' something or other, and your ... what else?' Bashir conceded but it was obvious he was irritated by all this.

'You talk as if I were trying to be self-sacrificing. One has to kill time somehow.' She laughed derisively, but there was a hint of sadness in her eyes.

'But, as I have stressed before, why do you not join the women's committee and work in a more structured manner? It is not as if you cannot get on with people. And in that way, you can gain influence, make your voice heard.'

'And what should I do then?' she asked, trying to sound innocent, but well aware of her husband's overpowering ambition. Even if she did something useful, it had to be trumpeted about, albeit discreetly, to gain approval. She bore his name.

'Zohra, what is one to do with you? Why can't you understand?' He sounded impatient. 'You fritter away your time, your money, your energy, in this way, while organized effort would bring recognition—prestige.'

'Will you try to understand if I tell you something?' she asked, although she was certain that he would not. As the expression in his eyes indicated an affirmative, she proceeded: 'I have started to feel that neither the time, nor the money, nor the energy, really belong to me. How can I work for what you call a reward? It has got to fit into the larger scheme of things …' She hesitated, then went on very quietly, 'I feel guilty whenever I have to give anything to the poor or the needy. I somehow start thinking what I should feel if, instead of giving I had to receive, and I feel ashamed that I should have this power to embarrass others.'

'I should have thought, that you would be proud of being able to help. Zohra, what *has* happened to you?' he continued, looking pained. 'Your old sense of values has gone. Even your ideas on religion have changed. Sometimes I even wonder if you are a Muslim any longer.'

'That is our difficulty.' Her eyes suddenly flashed. 'Because I am not fanatical about my religion, and am willing to respect other faiths, you jump to the conclusion that I am not a Muslim. Yes, in that sense I am not a Muslim.' She now looked at him steadily. 'I think it is true when they say, scratch a Muslim and he is a bigot underneath. But it is those who have a larger understanding who are the truly religious ones. You have little to do with Islam, but nowadays you always get worked up about it. I might as well ask, what has happened to *you*?' She leaned forward slightly in her chair.

'What do you mean?' he asked aggressively. 'I have always been a good Muslim, though I may not be strict about our religious observances.'

Zohra bit her lips to suppress a memory. She remembered the days when he had even talked irreverently of God.

'But,' continued Bashir, looking searchingly at her, 'you used to say your prayers regularly. Instead, now, you hold classes in the evenings, almost exactly at prayer time. You would never have done that before.'

'I know,' she said with quiet determination. 'Yes, my sense of values has undergone a change. I feel no urgency to pray now as I used to. After all, God is in our hearts, and I am sure He wants us to do what is right according to our own consciences, and not according to mere set rites. The more we think, the more individualistic becomes our faith. Besides …' she suddenly stopped.

'Yes, what besides?'

'Oh, I only wanted to say that prayers are after all selfish. It is because you want something for yourself, to save your soul or something; whilst ...' she was speaking awkwardly and might have trailed off into silence, but he would not let her.

'Yes, what?' he asked, determined to know what exactly had happened to her.

'While this is something useful,' she said almost fiercely, wanting to break away from this inquisition. 'No other time suited these women, and I would rather help them than sit and pray for my soul, or invoke God, for something that I want badly.'

As Bashir looked perplexed, she asked: 'But what is there for you to be so worried about? You never cared for prayers yourself.'

'No, if it were only the outward form you had discarded, I would not mind. But I feel it is your newly acquired political views that now influence your actions.' He sounded as if he held himself responsible for not having spent more time on guiding her thoughts. It irritated Zohra. 'It's people like Hamid who have done harm to Islam,' Bashir went on heatedly. 'Why did he publish that book? He calls it a novel. It's nothing but propaganda for his distorted political and social views.'

Zohra, who had only praise in her heart for Hamid's book, felt like crying out, 'It's people like you who have done the harm. It's only after politics brought religion to the fore that you even started thinking of yourself in terms of Islam.' But she made no retort. Her face looked strained.

Since Hamid's final release from gaol, he had again become the subject of Bashir's derision. Hamid had been released once, within a year of his being sentenced, when the viceroy—Lord Irwin—had prevailed upon Mahatma Gandhi to call off the non-violence movement and attend the Round Table Conference in London. Gandhiji had gone, hoping for a settlement, but had returned disappointed and given the call for a renewal of Satyagraha. Hamid, who had stayed in Bombay during the intervening months, was again sentenced to two years' imprisonment. He could have had his sentence reduced by paying a fine, but this was against the Congress creed. They would not in any circumstances aid an alien country to carry on its domination of India. The standing joke was: 'What fool would pay if he could be a free guest of the government?' Only the hospitality was such that it left many of them physically and emotionally maimed for a long time afterwards, and some even for life. Hamid was finally released after completing his full term. Apart from finishing his second novel in prison, he had written the book on rural arts he had wanted to write, and for which he had been collecting data.

Everyone now expected him to return to Hyderabad; but, instead, he wrote to say that as the Satyagraha movement was over, he would go to Shantiniketan, Tagore's 'Abode of Peace', and study the Model Village system, on which a number of students there were experimenting. He said this would be useful in Hyderabad, too. He had written to thank Khorshed and the others for having carried on with the bookshop against all odds, and expressed the hope that they would continue to do so and keep him informed of its progress. The publishing house was not doing well, but he said he would try to help them and maintain an interest in it from Shantiniketan. He wrote as if they were doing him a favour, while it was mainly on his capital that both these concerns were running. He never scrutinized the accounts. Hamid's staying away for a further period was a great disappointment to everyone. What Zohra did not know was that when Hamid had at last been released from prison, his health was shattered. The brutal hardships he had endured in prison had brought him to the edge of mental derangement. There was frustration all round, in the personal as well as the political sphere. He wanted to rush to Hyderabad—to Zohra, were it only to hold her hand, to pour out his soul, to look into her eyes. But the more he longed for this, the more the futility of it forced itself upon him. In desperation, he had decided to go to Shantiniketan in search of tranquility. He also desired to learn something, which would enable him to serve the people of Hyderabad.

Zohra had been counting each day, almost each hour, longing to see him again. Her deep wound opened up afresh and hurt more acutely. She knew it was because of her that he was not returning.

Time moved on; the children were growing up, and presenting their own problems. Bashir had little time to attend to them, but when he did intervene he took such stern measures that it only made the children live in greater fear of him.

II

It was the year 1935 and an epidemic of plague broke out in Hyderabad. Rats were dying everywhere. Many of the stricken localities were wholly or partially evacuated, the people moving away to camps erected on the outskirts of the city. The government was taking measures, but the disease continued to take a heavy toll. People were not accustomed to the seclusion that quarantine imposed on them and plague was a new disease here. 'Is this the fury of the God of the Whites?' they asked. The city wore a mournful air, with the dirge of funerals echoing in the empty streets.

It was hoped that Bashir, Zohra, the family, and the servants would escape infection since their house was fairly far away from the congested poorer areas where the disease raged.

Zohra's students could no longer come to the house. Distraught with the tales of suffering and deprivation, which were carried to the house by the servants, she donated money and tried her utmost to alleviate the suffering, desperate because she could do no more than this.

One day a little boy, one of her pupils, came weeping. 'My mother and sister are very ill,' he sobbed. 'Mother cannot speak.' He wiped his tears with his dirty torn shirt and then proceeded: 'Begum Sahiba, you are father and mother to us. For the sake of Allah, Begum Sahiba, you who are so merciful ... come and see my mother!' He repeated the phrases that he had heard his elders use.

'Have you informed the doctor in charge of your locality?' she asked, alarmed.

'No, Begum Sahiba.' He looked at her with frightened eyes. 'They say the doctors take them away and kill them.'

Bashir was not at home. Zohra did not tell her mother-in-law, for fear of being prevented from going, but ordered her car and driver and forthwith drove away with the boy. The one-roomed tenement was in a narrow lane. Zohra got out and followed the boy, stooping to pass through the low door into a semi-dark room. Inside, it was all filth and squalor. A woman was lying on a dirty mattress and she seemed to be unconscious. A girl of five or six lay beside her, tossing and groaning and weeping. An old woman sat by their side, trying to suppress the child's cries by frightening her with such awe-inspiring words as 'doctors' and 'police'.

As she saw Zohra, her old eyes became wide with fright. 'Who are you?' she demanded.

'Amma, don't be frightened, I have only come to try to help you, if I may,' said Zohra, bending down.

The woman looked up with horror in her eyes. 'Your car will attract attention ... That accursed boy ...' pointing to her grandson, 'he brought you here!' She started to talk in a low, squeaky voice. 'My daughter is lying with her eyes closed, half-dead. My granddaughter is as hot as burning coal, with fever. We have no one but Allah to help us. My daughter is a widow. Allah's curse has fallen on this already blighted house ... for what is a house, with its breadwinner gone? Begum, may you live long, but don't take my daughter away, where they kill them!' She could not weep for her tears had frozen.

But they welled up in Zohra's eyes. She tried to blink them away. Before this grim tragedy, her own, for the moment, seemed so insignificant. In the emotional upheaval that surged up in her, she forgot her fear of infection and, gently resting her hand on the grandmother's shoulder, sat down beside her. At this, the deadened heart of the old woman melted, and she started to plead, 'Oh, don't, don't take her away!'

'Amma, do not be alarmed,' said Zohra softly. 'I also have children. It is because I wish your daughter well that we must call in the doctor.'

It was only after Zohra had promised that she herself would see the daughter was properly taken care of that the old woman consented. Zohra wanted to fetch the doctor in her car herself but the old woman, now under the spell of her kindliness, would not let her leave. Poor woman, to her the danger of infection had little meaning. To her it was wrath or a cruel fate that had overtaken their city. Zohra had not the heart to wrench herself away as the old woman wailed: 'Oh, don't go ... please. Begum Sahiba ... don't go! You have come like an angel. People like you live always under the blessing of Allah. You have everything.'

There was a sharp pang in Zohra's heart.

'I am alone, please stay with me!' The old woman's voice quivered. Zohra sent the chauffeur in the car with an urgent message to the family doctor, while she herself waited in the dingy room. She did not know what comfort to give. There was nothing in that room that could be of any help. Soon the sick woman started to hiccough. In a few minutes, before they could realize it, she was dead. The old woman, her fear of the doctor gone, began to lament shrilly. The sick child's cries, too, became louder and more pathetic.

Sobbing, the mother asked Zohra to perform the death rites over a dying Muslim. Zohra closed the eyes and the mouth, placed her hands on the breast, said the appointed prayer in her ear, and was covering up the body with the dirty rag lying at her feet, when the doctor arrived.

Seeing Zohra in these surroundings, with the dead and the dying, the doctor lost his temper. He reprimanded her severely in English, so the others would not understand what he said. However, they understood his tone, but they were beyond caring. Zohra knew she had not been wise, but she also knew that she could have done nothing else.

The doctor took charge and hurriedly promised Zohra that he would attend to everything personally. While she was still trying to console the old woman, he unceremoniously led her to the car and almost pushed her in.

He gave her instructions for disinfecting herself. As Zohra leant back in the seat, tears were streaming down her cheeks.

When Bashir returned that evening and heard of what had happened, he was furious. 'What right had you to risk your life in this manner?'

'Yes, what right have I to do anything,' she answered, but her tone was subdued. The scene of squalor, disease, and death was still floating before her eyes. She knew he never had the same feeling for the poor. To him they were a race apart, although in theory he did take pride in an Islamic brotherhood.

'I think we had better call in the doctor and consult him about what you should do,' he said, rising to telephone. But catching hold of his hand she drew him down beside her, saying:

'Maybe, it is too late.' She laughed, trying to make light of what to her seemed unnecessary fuss. But, noticing how upset he looked, she continued: 'I have already followed the doctor's instructions. If in spite of this, something does happen, it will be just kismet!' But for the first time in many years, she looked at her husband with affection in her eyes.

'Zohra, you never stop to think. You are so impulsive.'

'You mean foolish.' She tried to brush away his fears. They were seated on the divan.

She placed her hand in his big square palm and nestled up to him.

Bashir was moved. Putting an arm around her shoulders he held her closely, and said: 'Zohra, I am lucky to have you, although at times I know I have made you unhappy. But I cannot have you expose yourself to such risks.' It required an effort to express such sentiments, but he continued: 'You have been looking so frail during these recent years that I feel anxious.' He looked at her with tenderness and longing.

'I am all right, really. It is only that I am made that way.' There was a catch in her throat. She wished she could respond more fully to her husband's love. But their spirits belonged to two different worlds, and a deep sensitive face, with ideals and faith akin to her own, forever disturbed her.

'You forget that I am much older now,' she added carelessly.

'How old? Thirty-two? You don't look more than twenty-two. I do not know how you keep so young. It is not flattering to me to have people take you for my daughter!'

'Nobody really does. They merely like to feed a woman's vanity.' She was wanting to humour him. But she had also somehow become sceptical of compliments about her own youthful looks.

Nevertheless, sitting side by side, the gulf between their ages seemed much wider than its twelve years. Bashir was now nearly bald, which gave his high forehead a massive look. He had put on considerable weight, too, during the last few years. Altogether, he looked smug and middle-aged besides Zohra's still girlish face and figure.

Zohra had sent money for the mother's and daughter's funerals. She had developed such a horror of unhappiness that she only felt thankful that the poor orphaned girl did not survive the illness. She arranged to have the son taken into their household as soon as his grandmother had left for her village, promising to educate him and to bring him up in their home with love and care.

About a week later, when the incident had somewhat faded from her memory, Zohra felt cold shivers running through her body. She called in the ayah and went to bed. Gradually the shivering subsided and her body began to burn. On the third day, all the symptoms were manifest, leaving no doubt about the illness. She was running a very high temperature. The family doctor was in almost constant attendance. Day and night nurses were engaged. Only a partial quarantine of the patient was practised. The children, however, were sent away to Safia's; in a fit of impulsive affection for her brother's children, she had offered to look after them. Elderly relations visited the house freely. In times like this, they seldom thought of their personal safety, they usually became more fatalistic. Even Bashir, who was more particular than the others, could not stop them from coming. Besides, with the grim spectre of death hovering over a young life, they could not be kept away on any account.

The nurses notwithstanding, there was always a member of the family in Zohra's room, for how could she be left alone in the care of professional women? Zohra's father, the Nawab Sahib, who was now bedridden, was unable to come to his daughter's bedside. But through dim eyes he read the Koran and prayed almost throughout the night. Zubaida Begum had arrived, but she could not approach Zohra save with tears streaming down her face. So she had to be kept away from her daughter as far as possible. Bashir's mother, however, was a more capable woman, but she was not a soothing influence. Mehrunnissa, apart from the consideration of carrying infection to her children, was worse than useless, complaining about her own heart and nerves. She had real cause for this, as her married life now seemed purposeless and empty. Bashir, though stricken with grief, had not the temperament needed to sit by a sickbed and humour a patient. Safia, whose antagonism to Zohra had become even more acute, excused herself on the apparently legitimate ground that she could take no risks, as she had all the children staying with her. But to her husband she said:

'One hates to say it, but it's the wrath of Allah on her.'

'Don't, don't talk in that manner! There are times when one must forget.' In spite of all his efforts to control his voice, there was a tremor in it.

'You are so generous, so kind-hearted, but to me when a thing is finished, it's finished. I'm nice to her children. I can do no more.' Safia's voice was bitter.

'It does not become a woman to be so hard,' said Yusuf, looking worked up, and hurriedly left the room. Safia's mother was greatly distressed to find her daughter still so indifferent and unmoved. For even when she sent a servant to ask after Zohra's health, it was more out of duty. On the other hand, Yusuf often dropped in to enquire, and altogether showed a greater warmth of feeling for the whole family.

The following day, however, old Unnie now almost eighty years old, being forbidden to go to Zohra's house, had, in an agony of feeling, gone to Safia's to see Zohra's children. As she was totally exhausted, Safia asked her to stay the night. Yusuf had not returned from a dinner at the mess, and Unnie, whose heart was bursting with Zohra and her illness, went and sat at the foot of Safia's bed, massaging her ankles and feet. Safia, curious to know all about Zohra's life in her mother's home, encouraged her to talk. Also, waiting for Yusuf was always dull, and Zohra's illness strangely obsessed her. Unnie went on with her rambling monologue of a conversation, during which she said:

'Our Nawab Sahib could not bear to see even a nail of hers hurt, and our Chhoti Bibi—may Allah give her healing—' said Unnie, wiping her eyes, 'was wholly devoted to him. She was content in her home, and was not like other girls who go to school in their carriages in order to be seen by strange young men through the carriage windows. Why, she would not even lift up her eyes to the boy-cousins who came to our house.' Unnie's tone in her present mood ranged from devotional fervour for Zohra, to contempt for all others, so that Safia could not help commenting:

'But, Unnie, she seemed to be quite at home with men at her bridegroom's home. She even entertained them while Bhai Jan was away. I don't say that there is any harm in that, but ...' her tone implied the opposite.

'*Ai-hai,*' Unnie exclaimed, 'they must have come for books. Our Bibi has a passion for books. Allah knows, she was very unhappy to give up her studies for marriage.'

'Maybe, Unnie, she was fond of one of her cousins,' Safia suggested pointedly: she was dying to know more about Zohra before her marriage.

'*Arrey, towba!*' Unnie expostulated unable to hide her indignation. 'Our Zohra Bibi—may Allah preserve her for a hundred and twenty-five years—was not like that. *Ai-hai*, a modest girl never thinks of marriage until she is married.'

'But that isn't quite natural—is it?' Safia tried to lend her question an innocent air.

'Begum, it is quite natural for girls like our Zohra Bibi. Modesty is the greatest ornament of youth, and our Bibi was the image of modesty!' Unnie was worked up now. With Zohra lying seriously ill, she could not brook one word that might suggest she was anything less than an angelic being.

'Why, when you say that men came to see her alone in her bridegroom's house, I can tell you that she even turned out your bridegroom. I heard her with my own ears ask him never to enter her private rooms again while she was alone, although he begged her a thousand times, with folded hands, to let him come.' Unnie was pouring out the story with righteous indignation, not stopping to think of its effect on Safia.

'What?' exclaimed Safia, her eyes starting out of their sockets, as her deepest doubts about Yusuf's culpability were confirmed. She sat up suddenly as if electrocuted. 'It must have been someone else!' she cried.

'*Ai-hai*, Begum, don't I know your bridegroom? He even talked to me. Allah preserve him; he has a kingly grace.' She remembered the notes thrust into her hands, and also the gallant manner in which he had treated her. 'May you, couple of the sun and the moon live for ever and ever!' Unnie, now alarmed, started to flatter Safia unreservedly, in an attempt to undo the harm she had done with her own thoughtless words. Looking at Safia, who sat as if mummified, encircling her knees with her arms, Unnie continued:

'He had only come to borrow some books. Begum, you know, our Zohra Bibi has so many of them, *owi*, she lives with books even more than men do.'

The defence was worse than the original disclosure; for Safia knew that Yusuf and books did not go together, especially the kind of books that Zohra possessed.

Yes, thought Safia, Yusuf had skilfully used the art of flattery on poor old Unnie, otherwise what could have kept her silent for so long? Safia jumped up from her bed, and without another word, staggered into her slippers and, while Unnie went on protesting, she walked away into the adjoining room visibly distressed.

Soon afterwards, bidding goodbye to the children, she asked the ayah to look after them for the night. Then she ordered the car and left for her mother's.

It was after midnight, and most of the household was asleep when Safia came rushing in. Her mother was still awake, squatting on her divan and reading passages from the Koran for Zohra's recovery. Safia dropped down beside her mother in a state of semi-collapse, flung her arms around her, and started to sob, her whole body shaking with emotion, as if she were in the grip of a fit.

'What is it, Daughter? Allah forbid, but what *is* the matter with you?' asked her mother bewildered, as fear gripped her that the dreaded disease might have overtaken her daughter's household too.

But for a long time, Safia could only utter monosyllables that brought no light to Masuma Begum. At last, through a flood of tears, she said:

'Amma Jan, your daughter-in-law is an angel!'

'I do not need you to tell me that!' said her mother, immediately on the defensive. 'But what sky has broken in upon you to suddenly bring you to see reason?'

'Amma Jan, I was deceived cruelly, wickedly deceived. How could I help suspecting her?'

'But how is it you have so unexpectedly recovered your senses?' asked her mother, as eyes unaccustomed to tears gradually began to be flooded by them. '*Ai-hai*, Safia, you have hurt her, perhaps mortally. Poor Dulhan never looked the same since you started your unpardonable behaviour. Even as your mother, how can I exonerate you?' She was wiping her eyes with her *pallau*.

'Amma Jan, I must go to her *now*, immediately, and ask her forgiveness,' said Safia, looking frantic.

'Daughter, are you mad? Do you know her condition?' Her mother's voice quivered. 'You would be the last person to bring peace to her.'

'Yes, yes, but I must ... I must ... I cannot wait!' said Safia impetuously. 'Besides, what if ...' She could not utter the hideous words. She started to cry again.

Masuma Begum now talked to her more gently; the years had mellowed her considerably. She herself needed comfort from the awful gloom enveloping their home. But Safia went on talking in an incomprehensible way, beating her breast from time to time. The only intelligible sentence she finally uttered was:

'Your son-in-law, I shall never set eyes on him again.'

Her mother felt as if she had been hit with a hammer. For a moment she was utterly dazed. Her son's home was near ruin; her daughter had now come declaring wildly that she had left her husband's house. Could it be that a wrathful Providence had marked out the lives of both her children simultaneously for destruction? As the frightful thought glared at her, she instantly pulled herself together and took command of the situation. She started to question Safia in a calm, sensible, and authoritative manner. She was able to get only a part of the story from her daughter's somewhat incoherent outpourings, and thus obtain some glimmer of the truth.

After she had drawn her own conclusions, Masuma Begum counselled Safia to have patience till the morning. She then allowed her to stay the night, although in view of the illness, she dreaded the idea. But there seemed to be no alternative, and leaves from the neem tree were kept burning in their room to act as a disinfectant.

When Yusuf arrived home in the early hours of the morning, he was in no condition to realize that his wife was not in the house. Lately, he had taken to drinking heavily. It was only in the morning, when he awoke with a bad headache, and was ordering coffee, that he discovered Safia's absence. No one in the house was able to throw light on his wife's sudden departure so late at night. He could only think of Zohra ... He had not the courage to send a servant and make enquiries. Dressing hurriedly and, gulping down some black coffee, he went straightaway to his mother-in-law's house. At the sight of him, Safia went into hysterics, and gradually Yusuf started to comprehend the situation. Since Zohra's illness, he had hardly been himself. Weak and vainglorious though he was, he was not really a vicious person, and the last few days had been spent in terrible mental torture. He had been wanting to make amends to Zohra, but did not know how. Mingled with very real concern was the superstitious fear that she might die without his being able to atone for his behaviour.

He now tried to talk to Safia, but she would not listen to him. Acute shame and remorse made her quite distraught. Masuma Begum took him aside, but before she could say anything, Yusuf poured out such bitter self-reproach that, instead of blaming him, she had to try to calm him down. The dashing army officer now cut a pitiable figure; pulling his hair, beating his head and altogether behaving in an ignoble manner. Masuma Begum squirmed inwardly, wishing he would show more dignity. She feared he might even have a fit. He cursed himself a thousand times for having talked so recklessly. In a state of frenzy he cried: 'I was mad! But Amma Jan, I was not responsible! Some devil had mounted my head.' He could not, of course, confess that, besides wounded vanity, the devil was jealousy at the sight of Zohra and Hamid apparently so engrossed in each other. However, his mother-in-law, with supreme self-control, showed far more understanding than Yusuf could have ever expected.

This, together with the strain of Zohra lying seriously ill, was enough to shatter the strongest of composures, but the Begum took complete command and, with great tact and strength of purpose, persuaded Safia to go back home with Yusuf. For, with her practical frame of mind, she knew that in that alone lay Safia's salvation. She herself had learnt hard lessons from life, and she knew that her daughter would have to overlook many weaknesses in her husband. Besides, Yusuf appeared keen not to break up the home. Partly it was deference to public opinion, and partly because of a genuine fondness for Safia. Masuma Begum, after long arguments, also convinced Safia that Zohra was in no condition for scenes. But she said that Safia could ask Bashir, who had also by now learnt what had happened, to give a message from her when he thought it opportune.

26

Zohra's condition worsened. The family doctor said: 'She still has a chance of pulling through, if only she would put up a better fight. Is there not someone who can rouse her to it?'

'Her whole family is here; they are all trying their best,' said Bashir, looking worn out.

Safia, on hearing this, could only exclaim spontaneously: 'Allah knows, Bhabi Jan has made her way into everyone's heart. Look at the servants. Look at Amma Jan. I know she would rather lose me, if it were a choice between us!' Safia was struggling to compose herself.

'I know her worth,' remarked Bashir drily.

'Amma Jan would lay down her life this moment if it could help in the smallest degree,' went on Safia. 'Unfortunately, she does not know how to soothe and comfort where it concerns so grave a situation.'

'But the question now is, what *is* to be done?' asked Bashir, with helpless impatience.

'I can only think of Hamid Bhai Jan,' said Safia impulsively. 'Although ordinarily he has a nervous disposition, he is very gentle and understanding and is able to give a great deal of strength to people who need it. It is his sensitive nature really. He will surely come if he can be of the least help.' Safia was anxious to do everything she would have previously rejected. She also longed to see her brother again.

'Yes,' agreed Bashir, not knowing what else could be done, 'perhaps Hamid can be of help. There is no harm in his coming.' However he could not help adding: 'Anyway, he has neglected his family long enough, and it is time he made a visit.' But Bashir knew that, for some reason, Hamid understood Zohra better than he did himself. In any case, the matter was too urgent for any sense of rivalry. He sent an express telegram asking him to return forthwith.

Hamid had not even been informed of Zohra's illness. On receiving the wire, he felt a world of darkness envelop him. He started for Hyderabad immediately.

By a strange coincidence, that very day Zohra, who despite her critical condition, was mostly conscious, said to her husband:

'I want to see your brother.'

'He will be coming soon.' Bashir's face looked tired and drawn. Zohra closed her eyes again.

Hamid was met on his arrival by Safia, who was waiting in a closed car at the station to receive him. After the anxious preliminary enquiries and replies, Hamid fell into a silence. But Safia, continuing to give details of Zohra's illness, suddenly burst into spasms of uncontrollable weeping. Hamid, whose head was in a whirl, was scarcely in the mood for such outbursts; he was afraid even to try to pacify her, lest he should lose control himself. Safia put both her hands on his shoulder, dropped her head on his chest, and through her sobs started to speak in short, disjointed sentences:

'I misjudged her ... What shall I do? Tell me, what shall I do? I'm not even allowed to go near her ... I hate him ... Yes, I hate him ... I want to die. Hamid Bhai Jan, ask her to forgive me. Promise me you'll do that! Otherwise my soul would suffer eternal damnation. She is a saint! ... Yes, now I know she is a saint, and you too!' Safia would not admit even to herself that she had been almost eager to misjudge Zohra, because she was jealous of Hamid's greater interest in her. Hamid could only make out fragments of what she was saying, but with so many other thoughts and feelings overwhelming him, he could not ask her any questions. He did not even know whether it was self-pity or concern for Zohra that dominated Safia's mind.

Hamid's arrival, which to his mother would ordinarily have been such a joyful event, was now only a more convincing proof of the very grave condition of her daughter-in-law. Also he looked gaunt and grey, still suffering from the results of his harsh imprisonment. This, together with her own daughter's behaviour, was wearing her down. Safia had gone back to her husband's home but, from the way she was behaving, no one could say what might happen next.

From the day Hamid arrived, Zohra began to improve. It seemed as if new life had been instilled into her. He spent long hours by her bedside. She would take nourishment and medicine from him more readily than from anyone else.

Whispered gossip started to circulate in the house, and the servants of the household, who were all so fond of both Zohra and Hamid, indulged in idle talk which was tinged with a wistfulness. They thought it a pity that Hamid had not married Zohra. Masuma Begum herself was stoical but the servants suspected that secretly she also thought as they did.

The house was thoroughly disinfected as Zohra's illness left her. But she was regaining her strength only very slowly.

'Hamid works magic,' Bashir admitted grudgingly to Safia. They were now involuntarily drawn towards each other, having found common comfort in talking about Zohra.

'Bhai Hamid has always had a soothing influence on people, but he has now acquired a new tranquillity. Haven't you noticed that?' asked Safia.

'Yes, I admit he is now calm and unruffled even when challenged about his strange unpractical ideas. But he now insists on taking charge of the family estates. After he has completed his course at Shantiniketan, he would like to go out and live in the villages. He is studying the Model Village System and feels he can successfully put the theories into practice here. I am only afraid we shall all go into bankruptcy, if he has his way. He really tries one's patience severely!' Then, changing this exasperating subject, he observed: 'But I am indeed grateful to him for the way he has looked after Zohra.' Safia could see that there was no suspicion in his mind.

As Zohra's health improved, she told Hamid of her life in Hyderabad since her return from Europe, and the great satisfaction she had gained from teaching poor children from the nearby slums. But above everything else, she seemed to be obsessed with the passion to discover the meaning of life. He tried to answer her as best he could, but she was never wholly satisfied. They talked for long hours. One day she said:

'Everyone has remarked on the way you've mastered your temper. Your entire demeanour has changed, and you seem to have gained an equilibrium. How did you do it?'

'Looks are often deceptive,' said Hamid wryly. 'But, anyway, as I've already told you, I often go to Gandhiji. In spite of all his preoccupations, he's always ready to listen and offer valuable advice. Zohra, you could meet him too. After all, one has to find a way of life,' he concluded wistfully, not daring to look at her.

Zohra knew what he meant, but neither of them had the courage to speak further about it. A silence descended over them both and enveloped them in its fold. At last, Hamid spoke as if resigned to whatever the future held for him: 'You know, Zohra, I now feel that death and resurrection are not necessarily achieved by physical death alone. A great suffering that shakes up one's entire being is a kind of slow, agonizing death, from which the soul is gradually resurrected, purified.'

Zohra gave a tired smile. She understood Hamid perfectly.

Gradually Hamid was persuaded to recount to Zohra some of his gaol experiences. She also asked him about his work at Shantiniketan. In turn, he asked her about her trip to Europe. As she told him of some of her experiences, he asked: 'But why did you not stay on? Paris would have done you a world of good.'

'It's just that I'm fated to cut short my holidays.' She tried to laugh but the recollection of Jacques distressed her still. Hesitantly, she tried to explain it all to Hamid. But it was impossible for her to relate it in any detail. Hinting at the fiasco of her last evening with Jacques, she hurriedly added: 'Looking back I feel it was my fault. I don't know how I could have acted so imprudently.'

Hamid could well comprehend the reason for that reckless behaviour, if reckless it really was. But what could he say? In spite of his newly discovered philosophy for living, the hopelessness of their position and the far-reaching effects of it, still troubled him and made him unhappy.

Hamid wondered if he should tell Zohra about his own sordid experiences in Calcutta after his release from prison and before going to Shantiniketan. Lonely and frustrated, he had found a thrill in mixing with an avant-garde artistic set. The women had found him charming and attractive, and, trying to get over his single-minded devotion for Zohra to the exclusion of all others, he had rushed headlong into a series of affairs. Apart from some excitement, he had found no solace, nor had he been able to accept that this was all life had to offer him. With one woman so deeply set in his heart, he had found it degrading to make love to another. Soon a revulsion for this empty way of living had set in. It was then that he had started visiting Mahatma Gandhi and holding long discussions with him.

Hamid, watching Zohra resting with her eyes closed, did not know how to explain or account for himself to a woman of her sensibilities. Besides, he was afraid of upsetting her in her weak condition. He put off the confession.

On one occasion, in a different mood, Zohra teased Hamid, saying: 'You're a quick-change artist!' She was sitting propped up in bed, reclining against the pillows at her back. Giving him a smile, she started counting on her fingers. 'Let's see. You returned from Europe in your English suit. Then as soon as you could, you got into a sherwani, the material of which, although hand-spun, was silk or *himru*. Soon afterwards, you took completely to rough cotton khaddar, but still it was a sherwani. And now ...' She cast a glance at him. He was wearing the new informal attire of young nationalist India. It was all in pure khaddar. The long kurta fell loosely down to the knees, over the trousers. It had a plain round neckline, with a buttoned opening on the left side. He wore sandals. He was not just then wearing the Gandhi cap which would have completed the outfit.

Hamid, in these simple clothes, presented an arresting figure, almost ascetic. To Zohra, he seemed to lend them an air of distinction.

Zohra continued: 'These changes of attire mark the different phases of your life. I approve of your wanting to identify yourself with our masses, although, with your background, it's difficult for you to become one of them.' She spoke as if she were regretfully stating a fact they all had to face. The gulf between the rich and the poor was so wide.

On another occasion, Zohra said:

'I was fearful of dying without seeing you again. It was only after you returned that I found peace.' Her eyes, which were now circled with dark rings, were mirrors of her innermost soul as they rested on him for a brief moment.

Hamid, looking into them, said: 'Zohra, you should have sent for me earlier. I didn't even know you were ill until I got that telegram.'

'I didn't have the courage. It was only when I felt desperate, and nothing seemed to matter any longer, that I could bring myself to ask for you.' He held her frail hand in his, but dared not speak. She continued: 'Without you I should have died. I was feeling so hopeless. I also wanted to ask you to look after my children. I want you to love them, to guide them.' She looked at him entreatingly.

'Do you have to tell me that?' His voice carried a gentle rebuke.

She gave him a grateful smile. Then she said tensely: 'I don't want Shahedah to marry the way I was married. Please guard her against that, in case ...' With her trembling hand she feebly pressed his.

Returning the pressure, Hamid released her hand, and answered: 'There's no need for you to concern yourself about these things.'

'No, I feel I must speak to you now. And ... there is one more favour that I would ask of you.' She hesitated, fidgeting with the green Kashmiri shawl thrown over her shoulders. Then quietly she said: 'Will you keep an eye on the youngest too?'

He could only nod assent; he wished she would not worry about all this at the present time. She leaned forward: 'Safia Apa loves him passionately, but I sometimes fear ... I don't know if we were wise ... I had hoped the child would make them more settled.' She was tense.

Taking her hand again, he pressed it more firmly to reassure her. He was seated on a chair by her bedside.

'The trouble with Safia is that she's so completely unbalanced. Even now, sometimes she says she would die if she has to continue to live with Bhai Yusuf; at other times she declares she would die if she has to live without him. What *is* one to do with her?' asked Hamid.

'They are both impulsive,' Zohra stated quietly, as if defending Safia.

'Yes, but Safia must learn to put up with certain things. Bhai Yusuf is very susceptible to good looks, and poor Safia has neither physical nor mental grace,' Hamid stated with obvious pain. 'I think she was born when the tension between our parents was most acute.' His voice was charged with feeling. 'But mother is wise about such things now. It is she who persuaded Safia to go back to her home. Bhai Yusuf also seems to have genuine affection for her; he wouldn't hurt her intentionally.'

'Yes, he is wayward, but kind-hearted,' agreed Zohra.

'Zohra, in spite of all the hints from the family, I don't yet know exactly what happened.'

'This last week, since my recovery, people have been inquisitive, trying to find out; but what can I say?' she said, looking as if the matter still lay heavy on her mind. 'But I've been wanting to tell you. Only I didn't know how to start on such a strange topic.' When Zohra had explained, Hamid was deeply shocked.

'To think that Bhai Yusuf was able to find ways and means of making advances, even in these semi-purdah surroundings, with prying eyes all around.' He shook his head in disbelief. 'The very fact that Unnie saw enough to tell Safia about it shows he was quite thoughtless and behaved without discretion.' Hamid did not know how to account for Yusuf's stupidity.

'It wasn't difficult,' replied Zohra softly, 'for he was keenly interested in Shameem. I suppose he longed for a child of his own. Shameem fascinated him and Bhai Yusuf often came to our house to watch him play.'

'Then?' asked Hamid, still sounding incredulous, for he knew Yusuf only too well.

'No, no!' protested Zohra unexpectedly, rising to Yusuf's defence. 'I think it was quite genuine.'

'So his interest in Shameem continued, did it?' Hamid asked, still with that sceptical look.

'Yes,' she answered, as if whatever else may have been, that one fact was indisputable. Then she continued with less certainty: 'But, gradually, he started confiding in me his difficulties with Safia Apa, and asked for advice. I was alarmed, and could only counsel him to be patient with her little weaknesses. What else could I say? I was young and incapable of offering guidance. I thought love could run a smooth course once it had been cemented by marriage. All it needed was effort.' She stopped as the hopelessness of her own present situation overwhelmed her.

A silence followed. Hamid, from his own state of mind, could well gauge her mental condition. But when he spoke, he only said: 'Let's get to the end of this tale!'

'There's little left to tell. Only, he started becoming more and more confiding. An intimacy developed in his manner, which I did not like—I felt uncomfortable. I was wondering what to do, when ...' She stopped abruptly.

'When ... what, Zohra?' Hamid was now anxious and curious to know how far his brother-in-law could go even with someone like Zohra.

'Oh, well!' she said hurriedly, wanting to put an end to the conversation.

'He started to talk wildly; he behaved strangely.' She could not proceed any further. It was not only the wild professions of love that she found embarrassing to repeat, but she still recoiled from the recollection of the utter lack of dignity and self-respect he had shown in his behaviour and pleadings.

Hamid saw that it was impossible for her to continue with the story. His face was drained as he said:

'I know I've no right, Zohra, to condemn others for being attracted to you. But, obviously, Bhai Yusuf behaved as he ought *not* to have behaved. Knowing you as he did, he should have shown a greater sensibility.'

Hamid was trying to curb his temper. He could not upset Zohra further in her present condition. But, looking at him, Zohra realized it, and sadly added:

'There must be something wrong with my behaviour that people should misunderstand me so easily.' She spoke earnestly. Since her illness, she had been trying to search her mind with complete honesty regarding her various actions. But where Yusuf had been concerned she knew it was not merely that she had not tried to attract him, but had desperately wanted to keep him at a distance, for his overfamiliarity had been distasteful to her.

Hamid knew only too well that Zohra's behaviour was not at fault. But he also knew that once he started enumerating her charms, he himself might start acting foolishly. He could only say:

'I could forgive him everything, but that his disappointment should have taken such a vile form of revenge!'

'In Ajanta, he saw what we ourselves tried not to see,' Zohra faltered.

'But how you could have kept quiet under his insinuations and Safia's attitude of self-virtue, is what I don't understand.' There was increasing admiration in his eyes. But Zohra was not looking at him.

'It was difficult,' she admitted, 'very difficult at times, especially when she kept Iqbal away from me.'

'But then, how could you?' he reiterated.

'How could I have disillusioned Safia Apa?' she asked simply. 'And perhaps, ruined her home? Besides, in spite of everything, I was fond of her. She is the sister of the two men with whom fate has got me so irrevocably entangled,' Zohra added with a wan smile.

'Zohra, you are so noble! I don't think I could have borne it, even though she is my own sister.' Hamid could say no more, but his eyes spoke volumes.

'You don't realize,' said Zohra, in a subdued voice. 'Love means so much more to a woman. You gave me a deeper understanding, made me aware of myself. Above all, my self-righteousness thawed under these uncontrollable emotions that you aroused in me. My debt of gratitude to you is profound and everlasting.' Then sadly she stated: 'I have a very deep feeling for your brother also, only it is all so different.' She could not proceed. Her hand passed over her tired brows.

'I know.' Hamid accepted the fact quietly. Then added: 'How often I have wished I had not brought this unhappiness into your life.'

'Please, don't say that again.' Her eyes gently upbraided him. 'I should willingly have suffered a thousand times over. You have given me my deepest faith,' she said fervently. Her face now looked strangely elevated. To hide her feelings, she could only close her eyes.

Hamid's elbows rested on the edge of the bed, as he held his bent head between his hands. Now that Zohra was gradually regaining her strength and life was returning to her, Hamid was troubled by this close proximity. He felt he could no longer even hold her hand to give her comfort. Spending so much time with her alone, always under a restraint, was torture.

Hamid also knew that people would no longer tolerate their being together on their own, now that she was on the way to recovery. He decided to return to Shantiniketan as soon as possible. But before leaving, he said: 'Zohra, if you ever really need me, you know you have only to let me know.' Zohra wept bitterly. This was her destiny.

Hamid had taken advantage of this visit to Hyderabad to see how the bookshop was working. But, as always, he was not able to understand the business side of it. He was also keenly interested in the progress of the publishing house. He spent long hours with his friends and former colleagues, including Khorshed, and planned further ways and means of helping them from Shantiniketan, for he realized that there was no way forward for Zohra and himself. Living under the same roof was impossible. It might be different when he could go away to work in the villages.

After Hamid's departure, Zohra's recovery suffered a setback, and gradually the doctors realized that she had developed pernicious anaemia. As month after month passed, she grew increasingly feeble. The plague had weakened her body, so there was no physical resistance against the emotional struggle within her. Hamid's presence had helped her to pull through the crisis, but his going away had left an even greater void in her life. Safia, the pendulum of her affections having swung back again, was a very frequent visitor. She tried her best to cheer Zohra up, and often brought Iqbal for Zohra to play with.

Zohra made an effort to hold on to life for the sake of her children. Hamid heard disquieting reports from his family but he did not know what he could do. Six months had passed since that attack of plague when news reached him that Zohra was entirely bedridden and that the doctors had even begun to despair for her life. In spite of every effort, her condition was worsening steadily. Allopathic, homeopathic, and ayurvedic, as well as unani remedies had all been tried out in turn. She usually responded to a new treatment favourably for the first few days; then again there was a setback.

Bashir, anxious and worried, said to Safia: 'I think we should send for Hamid. It is not as if he were really doing anything very important at Shantiniketan.' Bashir could never get over his contempt for the kind of work his brother was engaged in. 'But his influence over Zohra is certainly miraculous.'

Safia, whose affections and sympathies were again being generously lavished on Zohra and who, in her impulsive manner, was blaming everyone else for her own behaviour, thought to herself sarcastically, 'When she is well, she is Brother's affair, but when she is ill, she becomes poor Bhai Hamid's care!' Safia was now disposed to condone whatever Zohra and Hamid might have done. She would even have connived at an affair between the two if that were now possible.

Hamid came. He had aged considerably in the last six months. The furrows between his eyebrows had deepened. The grey streaks in his hair, especially at the temples, were more noticeable. In the first flush of happiness at seeing him, Zohra again improved slightly, but her general weakness and the sadness that enveloped their lives had robbed her of all illusions. She knew she was dying, so did Hamid. He found it an agony to watch her large eyes staring out of the dark deep wells of their sockets, and dwelling fondly, but still reticently, on him. How could he bear to sit and watch her dying by slow degrees? Even the plague had been more bearable.

Zohra's mind was still alert and clear, but everything else was failing. She often breathed in quick gasps. Apart from this there was no acute pain, only the inconveniences and little aches caused by weakness and being confined to bed.

One day, as Hamid entered, she asked the nurse to leave the room. Beckoning him close, so that her laboured breath almost touched his face, she said in a low but clear voice:

'You have been so good to me as only an angel could. How lucky I am to have you here. I want you now to recite whatever is sacred to you, that our spirits might be united for ever.' She gave a faint smile and placed her small hands in his. He saw the emaciated fingers, now slightly swollen, and the nails turned blue.

'What do you wish me to read? How can our spirits be more united?' he asked, in a voice drained of all emotion, gently brushing the little curls off her forehead. These last few days he had been lost in admiration of her silent courage in the face of death, and the smile of resignation with which she greeted people. It was as if all her strength were drawn into her spirit.

'Something that will give meaning to this parting,' she whispered.

A sob welled up in Hamid's throat and he closed his eyes to hide the turmoil. At last, clasping her hand in both of his, he slowly started to recite a verse from Kahlil Gibran's *The Prophet*.

For what is it to die but to stand in the wind and to melt in the sun?
And what is it to cease breathing but to free the breath from its restless tides,
that it may rise and expand and seek God unencumbered?

Only when you drink from the river of silence shall you indeed sing.
And when you have reached the mountain top, then you shall begin to climb.
And when the earth shall claim your limbs, then shall you truly dance.

She listened with eyes closed. When he stopped, she opened them again, and smilingly beckoned to him to come still closer. He bent down his tortured face and pressed his lips to her forehead, her eyes and her lips, as if touching something very sacred.

She looked so exquisite, so purified; he could not speak. She was calm with a spiritual resignation.

'Your spirit has been closer to me than anything I ever imagined. Mine, I know, will always be with you and the children,' she said slowly.

Suddenly he felt numb with terror. 'Zohra, my heart, you can still fight; you're young. Please, please, make an effort!' The panic of losing her almost made him choke.

She gently touched his sensitive face with her pale trembling hands.

'It's no use. I shall be freeing you also, by leaving you. I do not grieve for you. I don't grieve for him. I only grieve for the children. I tried to live for them, but I didn't have the strength to fight everything ...' She heaved a feeble sigh. 'They are in your care now, as well as their father's.'

It was amazing how coherently she could still speak. She closed her eyes. Hamid remained as if in a stupor. After a little while, Zohra, opening her eyes again, said:

'Promise me one thing.'

'What ... my beloved?' His lips quivered.

'First, promise,' she said, with the impatience of one who wanted to say all that she had to say, and who knew that every moment was precious.

'If it is in my power ...' He could not look at her.

'I want you to marry ... I want you to be happy.'

'Zohra! ...' he faltered, unable to go on.

'It will be easier to marry when I am gone. I know you will remember me. If there is an afterlife, we shall meet again. If our spirits are still close to each other, we shall be united there ...' She faltered, as her breath came heavy. Then proceeded: 'I have begun to dread waste—the futility of life—it gets one nowhere!' Though speaking in a laboured voice, her thoughts were well collected as if she had considered the whole matter through.

He was too dazed and bewildered to say anything. Nervously he kept passing his fingers through his hair.

She asked him to open the drawer of her desk, and to bring her a packet wrapped in mauve brocade that he would find there. When he had brought it, she said:

'Abba Jan gave this to me; it is now yours.' He was too overcome to utter a single word. But she seemed to understand. She motioned him to bend down and placed her hands upon his head as if in blessing; then, beckoning him close, she gently touched his forehead with her lips. She closed her eyes again. She was now wholly exhausted, both physically and emotionally. Hamid started to feel the ache of life return and, summoning the nurse, hurried out blindly.

Once alone in his room, he flung himself into the armchair. Suddenly his emotions, maintained under severe control for so long, were released in a torrent of wild grief. He wept distractedly, distraught with sobs that shook his entire being and seemed to echo through the room. He did not know how long it was before he came to himself. His mind instantly flashed back to his last moments with Zohra. It was as if during this time he was under an opiate. With the return of his senses he accused himself bitterly. What had he done? What good had this great love of his achieved, but to help kill the woman he loved? The future seemed so utterly dismal. He shuddered.

Yes, he thought bitterly, now he could come back and settle down in this house of mourning and be near her children. It was the only amends he could make. It was not only for her sake that he wished to be with the children; he had come to love them and, as far as possible, he would see that no harm ever came to them. She had asked him to see that Shahedah did not marry the way she herself had been married. He would also have to absorb himself in work to retain his sanity.

It was late at night, when they had all retired, that Hamid found the courage to take up Zohra's gift and unwrap the cover of mauve and gold brocade. Inside, he found a copy of Hafiz, the Persian Sufi poet, who had sung exquisite lyrics of love and wine, wherein love and wine are both divine, and union with God is achieved through the beloved. It was an old and priceless copy, in beautiful calligraphy, illustrated in the ornamental style of Persian miniatures. In a corner of the first page was inscribed in small letters, 'From Zohra'. There was also a notebook bound in leather, in which she had made a collection of her favourite poems and quotations—English, Urdu, and Persian—over many years. Hamid tried to glance through some of the pages. He turned to the last page. Written in a shaky hand, probably during her last illness, was inscribed a poem by Mirza Mohammed Rafi Sowda. He read:

> *Mourn not the death of anyone—O Sowda,*
> *Weep only for those who are dying to live.*

Turning over the pages, he saw poems new to him. They were poignant and sensitive. Hamid knew they were Zohra's own.

After that final meeting with Hamid, although she retained full consciousness to the end, Zohra spoke little. On the third day, in the presence of her family, including Hamid, her restless soul found peace.

If Zohra had one last conscious image of this world, it could only have been of the tortured man she loved, stooping with his arms round her sobbing daughter—slender and tall for her years. It appeared as if he was comforting her, and yet was strangely leaning on her for comfort himself.

It is often said that people look lovelier in death; for Zohra this was undeniably true. For it was a beauty that had been refined through sorrow and suffering.

Inna L'il—Lahi Wa Inna Ilayhi Rajioon.

Verily, to God do we belong, and to Him do we return.

Glossary

Abba	:	Father
Ai hai	:	An exclamation of mild disapproval or censure
Amma/Ammi	:	Mother; this can also be added to address an aunt to indicate closeness and endearment e.g. Chachi Amma
Apa	:	A form of address used for an elder sister, female cousin or older female friend
Arrey	:	An exclamation of surprise
Ayurvedic	:	An alternative system of medicine
Badi	:	Elder
Badi Amma	:	The first of a husband's four wives (permitted in Islam) is always the eldest and is addressed as Badi Amma by the children of the household
Badmash	:	Scoundrel
Ber	:	An Indian fruit resembling a berry
Bhabi	:	Brother's wife
Bhai	:	Brother
Bibi	:	A respectful term used largely by servants to address a younger female member of the family
Bismillah	:	*lit.* 'In the name of Allah (God)'. Muslims often use this word to preface what they say
Chacha	:	Father's brother
Chachi	:	Father's brother's wife
Champak	:	A highly scented white flower
Chiksa	:	A perfumed powder used mainly by brides
Chhoti	:	Small or young; used to define the younger or youngest female member of the family
Diwan-e-Aam	:	A public reception room
Dulhan	:	Bride. Often, a daughter-in-law is referred to as 'dulhan' throughout her life at the home of her parents-in-law

Dupatta	:	A veil or scarf draped over a woman's head and upper body
Eid	:	Muslim festival celebrating the end of Ramazan
Falsa	:	A small black berry-like fruit used to make sherbet and jam
Firangi	:	A mildly derogatory term for a foreigner
Ghalib	:	The renowned poet of Urdu and Persian verse
Ghazal	:	Lyrical love poetry traditionally sung as a ballad
Gulmohur	:	A tree with bunches of flame-red flowers
Hafiz	:	The distinguished Iranian Sufi poet whose verses were thought to be prophetic
Himru	:	A handloom cloth woven in a variety of designs
Holi	:	A festival to usher in the spring
houri	:	A beautiful woman one would expect to meet in Paradise
howdah	:	A seat for riding on the back of an elephant, usually having a canopy
Jan	:	A term of respect and endearment attached to the end of a name or form of address
Julwa	:	Revelation: refers to one of several arranged-marriage ceremonies at which the bride and bridegroom see one another in a mirror for the first time
Kashta	:	The fold of a sari worn in the Maharashtrian style and tucked at the back of the waist
Mamoo	:	Mother's brother
Manja	:	A bride's pre-marriage period of seclusion during which she wears only yellow clothes
Masha Allah	:	*lit.* 'Praise be to God'; also used as an exclamation
Mehr	:	A sum of money paid to the wife in case of a divorce
Mian	:	A formal address for men and boys, used largely by the older generation
Moharram	:	Islam's annual month of mourning to commemorate the martyrdom of the Prophet's grandson and his kinsmen
Moulvi	:	A religious scholar and teacher
Mubarak	:	The word used to offer congratulations
Mushaera	:	A poetry recitation (in the presence of an audience) at which each poet recites his best compositions
Mushata	:	The professional go-between (usually a woman) between families when arranging a marriage
Musnud	:	A seat made with a mattress and cushions
Nani	:	Grandmother (mother's mother)

Nana	:	Grandfather (mother's father)
Neem	:	An Indian tree, the leaves of which have healing properties
Nikah	:	The registration ceremony in a Muslim marriage
Owi	:	An exclamation of surprised disapproval
Paan	:	Betel leaf folded with nuts and chewed as a masticatory
Purdah	:	A curtain and hence a system used to seclude women from men
Pallau	:	The end of a sari which is thrown over the shoulder or used to cover the head
Qazi	:	An expert and interpreter of Muslim law
Sala	:	The brother is the 'sala' of his sister's husband
Salaam	:	A gesture of greeting or respect in Arabic-speaking and Muslim communities, consisting of a low bow of the head and body with the hand or fingers touching the forehead
Shaadi	:	Marriage
Shamiana	:	An awning, usually a large decorative one, put up in the garden for weddings or other important occasions
Shantiniketan (also Santiniketan)	:	*lit.* Abode of Peace. The institution set up by Rabindranath Tagore as an international centre for learning, with special emphasis on the revival of Indian arts and culture
Shariat	:	The Muslim code of religious law
Sherwani	:	A fitted knee-length coat worn by Indian men
Sitaphal	:	Custard Apple—An Indian fruit
Swadeshi	:	*lit.* 'Own Country'. The swadeshi movement of Indian resistance was initiated in 1905 and advocated an affirmation of Indian goods in order to support home industries; also refers to any goods made in India
Takht	:	A traditional Indian sofa, usually a large one, on which several people can sit or recline in comfort
Taziya	:	A mausoleum-like structure usually erected during the period of Mohurrum to commemorate the martyrdom of Imam Husain, the Prophet's grandson, and his companions. On the anniversary of the martyrdom, it is carried out in a street procession
Tikka	:	A mark (usually vermilion) worn on the forehead by married Hindu women
Towba	:	An exclamation of severe disapproval or disgust
Unani	:	An alternative system of medicine
Unnie	:	Or 'Unna': wet nurse

Appendix

Dedication to the Original Edition of *Zohra*

To my sisters:
Sameena, Aalia and Razia,
in happy memory of our home in Hyderabad.
The younger two are no more.

*'The sigh is for those flowers
That pass away in glory.'*

Foreword to the Original Edition of *Zohra*

I read the earlier chapters of *Zohra* some years ago in Bombay. They impressed me by their vividness, and by the picture they gave of the old Moslem society of Hyderabad—a society of which I chanced to see a little before it disappeared. I have now read the whole book, and what now impresses me is the character of the heroine. She is both convincing and charming, and thanks to her the book is not only an interesting document but a creative achievement.

E.M. Forster

13 August, 1951.
Cambridge, England.

Preface to the Original Edition of *Zohra*

It was in a foreign land that I started writing this story. Perhaps that accounts for it being in English – a language that I cannot handle with sufficient ease for confidence. Looking around and talking to people, I became increasingly conscious of how different that little world of ours, in Hyderabad, had been and I felt the urgency to record it, for owing to the passage of time it was fast disappearing.

I was in an especially fortunate position to do this for I belonged to it, and was yet apart from it, and could therefore take a somewhat detached view.

I have tried almost literally to translate part of the phraseology peculiar to Hyderabad and therefore the story should be read as a translation – maybe often an inadequate one. Amongst our people, fiction, and especially romantic fiction, is deprecated. But I have put aside my misgivings on this score as it is only through ordinary human emotions that one can best convey the lives of people. And no society, however strict or conventional regarding its women, is yet free from emotional entanglements.

The conflicting political ideals among the Indian Muslims also naturally find a place in this book through its main characters. No picture of those times can be complete without Mahatma Gandhi, his *Satyagraha* movement, and its impress upon the people.

Although first written several years ago this story has been revised at intervals. There are many who have given me their kind help and criticism, and if I do not give a long list of names, it is only because it would make tedious reading. Nevertheless there are some whom I must mention. It was Rasna and Phiroze Mistry who not only helped to weed out much of the unnecessary detail, but were also helpful in re-shaping it and making it more presentable. I then showed it to Mr. E.M. Forster. His criticism was both encouraging and heartening. He pointed out, besides, some of the blatant flaws, in a manner which was most charming in a novelist of his eminence. He let fall ideas in the form of mere suggestions. I realise now how valuable those suggestions were, and how fortunate I was to come into contact with a great artist with so complete an understanding of other people's limitations.

Finally, it was Mr. K.P.S. Menon who, in his always genuinely helpful manner, at once consented to write the Introduction. This, notwithstanding the pressure on his time, was indeed gracious of him. There are two other names that cannot be passed over – Mr Ashton and Mr Chatterton whose spontaneous help was most welcome.

I here offer my warmest thanks to all those who have so kindly and generously given me their help.

ZEENUTH FUTEHALLY
6 August, 1951
'Sahil'
Pali Hill, Bombay.

Introduction to the Original Edition of *Zohra*

In his Introduction to my book, *Delhi-Chungking*, Pandit Jawaharlal Nehru said that since he became Foreign Minister he hardly ever had time to read a book. Regretfully he would look at shelves filled with books, sometimes take out a book and handle it with affection and then put it back again. Though the Foreign Secretary's duties are but hobbies as compared with those of the Foreign Minister, who is also Prime Minister, I too find, for the first time in my life, that I have no time to read anything less bleak than notes and drafts, telegrams and savingrams, and reports and despatches. The only exception in recent months was *Zohra*.

Zohra came in manuscript into my life a few months ago and I spent a delightful weekend with her. Fortunately it was a weekend free from the usual round of diplomatic parties. Even my wife was away: she had gone on a spiritual quest to Badrinath leaving me to attend to the secular duties of a secular state. The creator of Zohra herself, having just introduced her to me, retreated discreetly into the hills of Mussoorie. So I had Zohra all to myself; and she, me.

I could not have wished for amore genial companion. Zohra was charming, amusing, gripping and pathetic. Pathetic, because, like many a young girl of her generation, she was

> Wandering 'twixt two worlds
> The one dead, the other yet unborn.

One world, a world of dignity and decorum, scents and sherbets, Nawabs and nautch girls was dying. Into it was breaking another world with different values, vital, iconoclastic, antinomian. Zohra belonged physically to the former and spiritually to the latter and was crushed between both. A moving story, simply, sensitively and, in places, poignantly told.

I hope many a reader in India and abroad will enjoy a weekend with Zohra as I have done.

K.P.S. Menon
14 August, 1950
New Delhi